Applied Multidimensional Systems Theory

Nirmal K. Bose

Applied Multidimensional Systems Theory

Second Edition

Springer

Nirmal K. Bose (Deceased)
Electrical Engineering Department
Pennsylvania State University
State College, PA, USA

First edition published by Van Nostrand Reinhold (October 1981) ISBN-10: 0442272146;
ISBN-13: 978-0442272142

ISBN 978-3-319-83595-2   ISBN 978-3-319-46825-9   (eBook)
DOI 10.1007/978-3-319-46825-9

Printed on acid-free paper

This Springer imprint is published by Springer Nature
The registered company is Springer International Publishing AG
The registered company address is: Gewerbestrasse 11, 6330 Cham, Switzerland

Preface

Throughout his career, Dr. Nirmal K. Bose, a world-renowned expert whose research was heavily focused on multidimensional signals and systems theory, conducted a great deal of research and published many articles on high-resolution reconstruction of blurred and noisy images and processing of noisy images [1]. In 1983, Dr. Bose published a pioneering book entitled *Applied Multidimensional Systems Theory* [2]. The first edition of his book emphasized new research results that had been emerging over the previous decade, where considerable activities had been witnessed in the area of multidimensional systems theory, motivated by the variety of applications embracing multidimensional signal processing (M-D DSP), variable-parameter and lumped-distributed network synthesis, stiff differential systems, and continuous as well as discrete nonlinear systems characterized via the Volterra series.

At the time the first edition of this book was published, the need to introduce mathematical and computational tools supporting this subject matter in graduate curricula of most universities was becoming very apparent. Because no textbooks were available on M-D DSP in the 1980s, lectures in seminar courses, workshops, and continuing education offerings had been organized from a couple of edited books and a variety of research papers published in scattered journals. The emphasis at that time was based on the documentation of results considered to be of a fundamental nature leading to interdisciplinary applications in several areas of engineering, mathematics, and science. Both the progress that was being made and the existing difficulties in extending or adapting established one-dimensional techniques to the multidimensional situation were emphasized.

The fact that progress in the area of multidimensional systems theory was dependent to a great extent on the interaction between mathematicians, computer scientists, and engineers was emphasized as a reader progresses through the book. It is also noteworthy that soon after the first edition of Dr. Bose's book was published in 1982, a new textbook was published in 1984 by Dudgeon and Mersereau [3] emphasizing algorithms and hardware structures to use rapidly emerging multidimensional digital signal processing. Many graduate instructors used both of these textbooks in order to combine a detailed mathematical background with the rapidly evolving use of 2-D DSP technologies in many high-tech areas such as X-ray imaging, MR imaging, radar imaging, and ultrasound imaging.

In 2009, Dr. Nirmal Bose took his sabbatical leave from Penn State University and spent a great deal of his time developing the second edition of *Applied Multidimensional Systems Theory*. He worked closely with his M.S. student, Mr. Umamahesh Srinivas, who assisted him with his editorial revisions and entered all of the revised materials into LaTeX computer files. During the fall semester of 2009, Umamahesh's interactions with Nirmal Bose led him to the conclusion that the manuscripts for the second edition of the revised book were moving toward completion, although at that time Dr. Bose had not yet signed a contract with a publisher for his revised edition. Unfortunately, in November of 2009, Dr. Nirmal Bose passed away suddenly due to an unexpected heart attack, so at that time the new edition of his book was not published. This current second edition entitled *Applied Multidimensional Systems Theory* is a moderately edited revision of his second edition manuscript. Dr. Nirmal Bose has been retained as the primary author, and the book is being published in honor of his lifelong career accomplishments.

When revising the original manuscript to produce the second edition, Dr. Bose made a major effort to remove some of the detailed mathematical theory that was not particularly useful to students and instructors using the textbook for graduate courses, to revise the presentation of portions of the theory to make it more readable for students, and to introduce some new topics that were emerging as multidimensional DSP topics in the interdisciplinary fields of image processing. In this second addition, Chaps. 1 and 2 introduce essentially the same materials that were presented in the first edition, although considerable revisions were included to make the materials more accessible to students. Also, Dr. Bose inserted the new topic of "Gröbner Bases" which is now included in this edition. The second edition contains a new Chap. 3 entitled "Multidimensional Sampling," much of which was contained in the first edition but was not highlighted as a special chapter. The third and fourth chapters of the first edition have been combined into a single Chap. 4 that has been renamed "Multidimensional Digital Recursibility and Stability."

In the new edition, Chap. 5, entitled "2-D FIR Filters, Linear Prediction, and 2-D IIR Filters," places a considerable amount of emphasis on 2-D digital filter design to lead students to more efficiently use the mathematical theory underlying the practical areas of 2-D DSP filtering applications. Much of this material was included in the first edition's Chap. 6 entitled "Additional Applications," although the second edition highlights these concepts in much more practical and useful ways. Finally, in the second edition, the new Chap. 6 entitled "Wavelets and Filter Banks" is an important new material not previously published. This is a reflection on the fact that wavelet theory emerged from mathematical communities in the 1990s and beyond and is now a central portion of multidimensional DSP theory and applications.

The Editorial Committee established to oversee the publication of the second edition consists of Umamahesh Srinivas, Apple Inc., Constantino Lagoa, Pennsylvania State University; and Kenneth Jenkins, Pennsylvania State University.

In particular, the Preface was written and edited by the entire Editorial Committee, and the committee members' editorial efforts were focused on the six chapters. As mentioned previously, Dr. Umamahesh Srinivas was Nirmal's M.S. student at Pennsylvania State University who worked on the second edition manuscript that was near completion at the time of Nirmal's passing. Dr. Constantino Lagoa is a professor of electrical engineering at Penn State who spent many days during his early career working with Nirmal in the Departmental Area Committee on Circuits, Communications, and Control and taught many of the same courses that Nirmal taught over the years. Constantino's area of expertise is focused in the Control Systems Area, so his point of view is strongly connected to multidimensional signals and system theory, but his perspective of Nirmal's research and teaching was from a different technical direction.

The first time Kenneth Jenkins met Dr. Bose was in Tokyo, Japan, in 1979 where they both participated in the International Symposium on Circuits and Systems. It was memorable that Nirmal Bose led a small group of CAS attendees to an Indian restaurant so they could sample the Japanese version of Indian cuisine. The next time that Kenneth and Nirmal met was when they worked together on a conference program committee in Philadelphia in 1987. During this era, Kenneth began teaching a portion of a graduate course at the University of Illinois from the first edition book *Applied Multidimensional Systems Theory*. Then in 1999, when Dr. Jenkins joined Penn State as the Department Head of Electrical Engineering, he learned all the details of Dr. Bose's array of international activities, including delivering plenary lectures around the world, traveling to Germany for several summers on Humbolt Fellowships, and working with a broad array of international graduate students. He was an international scholar who brought worldwide visibility to the Electrical Engineering Department at Penn State University.

Dr. Bose was a truly international scholar. Born in Calcutta (presently Kolkata), India, on August 19, 1940, Nirmal Bose received a B.Tech. degree from IIT Kharagpur, India, in 1961. Thereafter, he traveled to Cornell University, Ithaca, New York, for graduate education, where he received a master's degree in electrical engineering in 1963, and then in 1967, he received a Ph.D. degree at Syracuse University. After a short stint at Princeton, he then went to the University of Pittsburgh as an assistant professor and rose to full professor of electrical engineering and later held a joint appointment with the Department of Mathematics and Statistics. At Pittsburgh he first wrote papers that showed his interests outside the traditional domain of circuit theory and in broader areas of system theory and mathematics. It was during this period that ideas for multidimensional systems slowly began to germinate, and during this time, he pursued interests in discrete mathematics, graph theory and related routing, and layout problems of large-scale integrated circuits.

Since the field of digital signal processing had come of age in the 1980s and faculty members at many universities were teaching courses on the topic, he consolidated his class notes in the undergraduate (senior)-level textbook *Digital Filters: Theory and Applications*, published by North-Holland Elsevier, NY, 1985 [4]. Around this point of time, the field of multidimensional systems and signal processing began to become somewhat diverse, and the many topics in which he published with students and colleagues are multidimensional approximation theory, model reduction, and filter design. In April 1991, he guest edited a second special issue of the *Proceedings of the IEEE* on "Multidimensional Signal Processing" in order to assess the status of the field since publication of the first *Proceedings of the IEEE* special issue on the topic in 1977.

In 1986, Dr. Bose moved from the University of Pittsburgh to Pennsylvania State University, where over the years he held several endowed professor positions. Here, at first, he devoted considerable time in developing laboratory and curriculum in digital signal processing in the EE Department with the help of younger colleagues. He also turned his attention toward more practical areas. An example of the latter was his interest in the burgeoning field of neural networks in which he taught courses, gave plenary lectures at international conferences, and was the principal author of a book entitled *Neural Network Fundamentals with Graphs, Algorithms and Applications*, coauthored with P. Liang and published by McGraw-Hill Book Company, NY, in 1996 [5]. Also in the early 1990s, Dr. Bose worked with Kluwer Academic Publishers to create a new journal entitled *Multidimensional Systems and Signal Processing*, for which he served as the founding editor-in-chief.

Dr. Bose authored, coauthored, or edited 15 books, published special issues of several journals in engineering and mathematical disciplines, contributed about 25 chapters in edited books, and authored or coauthored more than 150 journal papers. He was an elected fellow of the IEEE, served in various positions in IEEE Circuits and Systems Society including serving on the CAS Board of Governors, served in the editorial board of the *Transactions*, and served as the chairperson for its education committee. He was the recipient of the IEEE third Millennium Award in 2000, and in 2007 the IEEE Circuits and Systems Society honored him with the CAS Society Education Award. A list of Ph.D. dissertations supervised by him until 2006 shows that he had advised 30 Ph.D. students in addition to many more M.S. thesis students. He also served as an education advisor to the government of India for the United Nations.

On the personal side, Dr. Bose was a kind and gentle person who had high ethical standards from which he never deviated. Lastly, it is important to highlight that Dr. Bose was an intellectual giant in his field. His work was highly published and highly referenced in the literature, and he was highly recognized by numerous awards and honors bestowed upon him by his profession. It is in his honor that the second edition of this textbook is being published, and it is with great respect that the Editorial Committee would like to thank the Bose family for all the support they have provided to make this honorable publication possible. The views expressed in this book are Dr. Bose's own and the editors do not take responsibility for them nor do they assume full responsibility for any errors and omissions.

Editorial Committee Members:

Dr. W. Kenneth Jenkins
Chair of Editorial Committee
School of Electrical Engineering and Computer Science
Pennsylvania State University

Dr. Constantino Lagoa
Editorial Committee Member
School of Electrical Engineering and Computer Science
Pennsylvania State University

Dr. Umamahesh Srinivas
Editorial Committee Member
Apple, Inc.

Contents

Chapter 1
Multivariate Polynomial Fundamentals for Multidimensional Systems

1.1 Introduction

The subject of multidimensional systems is concerned with a mathematical framework for tackling a broad range of paradigms whose analysis or synthesis require the use of functions and polynomials in several complex variables. Its applications, which may range from the processing of spatial and temporal signals of diverse physical origin to the design of linear discrete multidimensional control systems, are already plentiful. The areas of image processing, linear multipass processes, iterative learning control systems, lumped-distributed network synthesis, nonlinear system analysis via multidimensional transforms and geophysical signal processing have benefited from the tools available in the theory of multidimensional systems [2, 6]. Progress towards the use of the theory in problems of very recent origin like multidimensional convolutional coding for communications has been increasing at a rate which is becoming increasingly difficult to track because of the wildly scattered nature of the voluminous publications by researchers from several disciplines, who are contributing to this area. This book aims to promote interaction between a broad spectrum of scientists and engineers so that not only are theoretical results developed to their fullest possible extent but also clear exposition and interpretation are provided for these results to become useful to practitioners in distinct but related disciplines. The presentation is intended to be concise but complete.

Research that had been started and conducted in the areas of multivariate network realizability theory (since about 1960), two-dimensional digital filters (since about 1970) and multidimensional transform analysis of nonlinear systems representable by Volterra series that outdates the preceding two areas just cited was included within a framework that was christened *multidimensional systems* in June 1977, when a Special Issue, guest edited by the author, was published by The Proceedings of the IEEE. A fertile arena for application of the developed theoretical results in multidimensional systems is multidimensional signal, image, and video processing.

© Springer International Publishing AG 2017
N.K. Bose, *Applied Multidimensional Systems Theory*,
DOI 10.1007/978-3-319-46825-9_1

Another Special Issue of The Proceedings of the IEEE was guest edited by this author in April 1990 and this one was devoted exclusively to the topic of multidimensional signal processing. The reader may wish to read the opening paper [7] in that Special Issue to grasp the fundamental limitations as well as scopes for generalizations of one dimensional signal processing theory in various spatio-temporal signal processing applications.

Some of the fundamental difficulties which either complicate the development of resources for meeting the needs of diverse applications or are responsible for limiting the scopes of attempts at naive generalizations of one-dimensional (1-D) results to the corresponding n-dimensional (n-D) situations have been recognized during the past several decades. Considerable analytical resources are required to go from 1-D to 2-D results and even after the associated bottlenecks are circumvented, the subsequent generalization from 2-D to 3-D may not be routine. While discussing the n-D real Euclidean space R^n, Daniel Asimov [8] correctly states that "strictly on their own merits, higher-dimensional spaces tend to blur together into multidimensional sameness." He goes on to remark that "it is often among low-dimensional spaces that the most dramatic transitions take place: as the number of dimensions rises, fundamental properties suddenly flash into existence or vanish forever, never to change again." This remark is dramatically illustrated by the celebrated conjecture made in 1904 by the distinguished French mathematician, Henri Poincaré, on a possible simple test to classify all three-dimensional manifolds. Though this conjecture remains unsettled to this day, its generalization was first solved in the case of n-dimensional manifolds in 1960 for $n \geq 5$ and in 1982 for $n = 4$ by ingenious innovative methods. The reader will encounter similar twists and turns in the journey through increasing dimensions within the scope and framework of the topics in this text. The undeniable challenge and the promised reward have propelled the subject-matter to a state of maturity that guarantees a continuing proliferation of increasingly complex and diversified nature of activities in the area.

It is important that the reader be exposed to the fundamental distinguishing mathematical features whose frequent deployment in the later chapters are anticipated. This will provide the background for comprehending the specialized results introduced and discussed as needed in later chapters. Every attempt will be made to strike a judicious balance between brevity and clarity of exposition. Where necessary, the understanding of a complicated concept will be facilitated by carefully selected illustrative examples. Nevertheless, the reader is expected to be capable of attaining a level of mathematical maturity possessed by first-year graduate students in algebra and analysis.

1.2 Multivariate Polynomials, Ideals, and Varieties

The reader is referred to [9] for the development of the theory of algebraic curves from introductory modern algebraic geometry, which contains in a more general but abstract setting the relationships between polynomial ideals and affine varieties presented here as relevant prerequisites for material in the subsequent sections and chapters of this book. Let R be a commutative ring with identity. Let $R[z_1, z_2, \ldots, z_n]$ denote the **polynomial ring in n variables** $z_1, z_2, \ldots, z_n$ with coefficients over R. This totality of polynomials in $z_1, z_2, \ldots, z_n$ over R forms a commutative ring with respect to the operations of ordinary multiplication and addition. The ring $R[z_1, z_2, \ldots, z_{(n-1)}][z_n]$, denotes the commutative ring obtained by adjoining z_n to the commutative ring $R[z_1, z_2, \ldots, z_{(n-1)}]$. Each element of $R[z_1, z_2, \ldots, z_{(n-1)}][z_n]$ is said to be a n-variate polynomial expressed in *recursive canonical form* in the main variable z_n whose coefficients are $(n-1) - variate$ polynomials from the ring $R[z_1, z_2, \ldots, z_{(n-1)}]$. Any element of the ring

$$R[z_1, z_2, \ldots, z_n] = R[z_1, z_2, \ldots, z_{(n-1)}][z_n]$$

is a finite sum of **terms** of the generic form, $a_{k_1, k_2, \ldots, k_n} \, z_1^{k_1} z_2^{k_2} \ldots z_n^{k_n}$, where $k_1, k_2, \ldots, k_n$ are nonnegative integers and $a_{k_1, k_2, \ldots, k_n}$ is the **coefficient** multiplying the **monomial** $z_1^{k_1} z_2^{k_2} \ldots z_n^{k_n}$ of degree $k_1 + k_2 + \ldots + k_n$. The greatest degree of the monomials in a polynomial is called the **total degree** of the polynomial. A polynomial, each of whose monomials is of the same degree, is called a **homogeneous polynomial** or, synonymously, a **form**. Any polynomial can be uniquely expressed as a finite some of forms whose degrees range from zero up to the total degree for the polynomial.

Example 1.1. Let $R = \mathbf{Z}$, the commutative ring of integers. Consider the polynomial, $a(z_1, z_2, z_3) \in \mathbf{Z}[z_1, z_2, z_3]$, specified to be

$$a(z_1, z_2, z_3) = z_1^3 z_2^3 z_3^3 + 9 z_1^3 z_2^3 z_3 - z_1^2 z_2^3 z_3^3 + 8 z_1 z_2^2 z_3^3 + 1.$$

As an element of $\mathbf{Z}[z_1, z_2][z_3]$, $a(z_1, z_2, z_3)$ is

$$a(z_1, z_2, z_3) = (z_1^3 z_2^3 - z_1^2 z_2^3 + 8 z_1 z_2^2) z_3^3 + 9(z_1^3 z_2^3) z_3 + 1.$$

On the other hand, $a(z_1, z_2, z_3)$, viewed as an element of $\mathbf{Z}[z_1][z_2, z_3]$, is

$$a(z_1, z_2, z_3) = (z_1^3 - z_1^2) z_2^3 z_3^3 + (9 z_1^3) z_2^3 z_3 + (8 z_1) z_2^2 z_3^3 + 1.$$

Next, suppose that the ring R contains no zero-divisors, i.e., R is an integral domain. Then, $R[z_1, z_2, \ldots, z_n]$ is also an integral domain (or domain). Furthermore, the total degree of the product of two non-zero polynomials from $R[z_1, z_2, \ldots, z_n]$ is the sum of the total degrees of each polynomial if R is an integral domain. Also, in that case every unit of $R[z_1, z_2, \ldots, z_n]$ is a constant and a unit of R. Suppose that the integral domain R is also an *unique factorization domain* (UFD) so that any element belonging to it is either an unit or has an unique representation (up to units) as a product of primes, where each prime belongs to the UFD. Then $R[z_1, z_2, \ldots, z_n]$ is also an UFD, whose primes are referred to as *irreducible polynomials*. A set of elements of an UFD is said to be *relatively prime* if no prime in the UFD divides all of them. A polynomial over an UFD is *primitive* if its coefficients in the chosen representation are relatively prime. A product of primitive polynomials is primitive. Furthermore, a primitive polynomial is a **prime** polynomial or **irreducible** if it is not divisible by any nonconstant polynomial of lower total degree. The integral domain R is called a **field** if every element of R other than 0 is invertible. If R is a field then $R[z_1, z_2, \ldots, z_n]$ is an UFD. A polynomial belonging to such an UFD can be uniquely factored, as a product of a finite number of irreducible or prime polynomials and an element of K.

A n-variate polynomial, $a(z_1, z_2, \ldots, z_n)$ in the n complex variables $z_1, z_2, \ldots, z_n$ will now be assumed to have its coefficients in an arbitrary but fixed field K. Each of the irreducible factors is guaranteed to be of degree 1 when $n = 1$ and **K** is the algebraically closed field of complex numbers; in that situation a polynomial of degree d has exactly d factors, a fact referred to as the "fundamental theorem of algebra." In the complementary situation, the irreducible factors may not be of total degree 1. For example, when **K** is the finite field Z_q of integers $0, 1, \ldots, (q - 1)$ for a positive prime integer q, there exists irreducible univariate polynomials of any arbitrary but fixed degree. This fact is used to advantage in the construction of algebraic codes. When $n > 1$, it has been shown that the set of reducible polynomials has measure zero and almost any multivariate polynomial is irreducible.

Let **K** be any field and $z_1^0, z_2^0, \ldots, z_n^0$ be a set of n-tuples of elements of **K**, and let $a(z_1^0, z_2^0, \ldots, z_n^0)$ denote the element of **K** obtained by substitution of $z_1^0, z_2^0, \ldots, z_n^0$ for $z_1, z_2, \ldots, z_n$ in a polynomial $a(z_1, z_2, \ldots, z_n)$. If $a(z_1^0, z_2^0, \ldots, z_n^0) = 0$, then the $n - tuple$, $(z_1^0, z_2^0, \ldots, z_n^0)$, is called a **zero** of the polynomial $a(z_1, z_2, \ldots, z_n)$. The set of all zeros defines the **zero-set**, which is an **(affine) algebraic set** associated with $a(z_1, z_2, \ldots, z_n)$. More generally, the intersection and finite union of any collection of algebraic sets are algebraic sets. An algebraic set may be the union of several smaller algebraic sets; otherwise, it is called **irreducible.** In fact, any algebraic set is expressible as a finite union of irreducible algebraic sets. An irreducible (affine) algebraic set is called an *(affine) variety*.

Definition 1.1. A real (or complex) algebraic **variety** is a point set V(S) in real *n-space* $\mathbf{R}^n$ (or complex *n-space* $\mathbf{C}^n$) of the common zeros of a set S of polynomials with coefficients in $\mathbf{R}$ (or $\mathbf{C}$).

Example 1.2. Consider the set of m polynomials, each of degree k.

$$f_i(z_1, z_2, \ldots, z_m) = (z_i - 1) \ldots (z_i - k), \ i = 1, \ldots, m.$$

The above set of polynomials defines a zero-dimensional variety of precisely k^m points.

Let $A \subset R$, where, as before, R is a commutative ring with identity. Then the set of all elements of R each of which is expressible as a sum $\sum a_k r_k$, where $a_k \in A$ and $r_k \in R$, is an ideal. When the number of elements in A is finite, the ideal so generated is said to be **finitely generated** with the a_i^s as basis. When the ideal is generated by a single element of A, it is said to be a *principal ideal*. When $\mathbf{K}$ is a field, every ideal I in $\mathbf{K}[z]$ is principal and the greatest common divisor of I generates I. An ideal I of R is proper if $I \neq R$ and a proper ideal I is maximal if it is not contained in any larger proper ideal. The residue class ring of R modulo I is written R/I and this is a set of equivalence classes of elements in R with reference to I.

Definition 1.2. An ideal R_1 of the ring R is **prime** (**primary**) if, for any $r_1 \in R$, $r_2 \in R$, the conditions $r_1 r_2 \in R_1$, and $r_1 \notin R_1$, imply r_2 (some positive power of r_2) is in R_1.

A principal ideal is prime if and only if the unique generating element is irreducible. A proper ideal I in R is prime (**maximal**) if and only if R/I is an integral domain (field). A primary ideal need not be a power of a prime ideal to which it belongs. Also, a power of a prime ideal need not be primary (this fact is used in a later chapter to show the invalidity, in general, of polynomial matrix primitive factorization when the number of independent complex variables is greater than 2). Let I denote the ideal generated by any set S of multivariate polynomials in $\mathbf{K}[z_1, z_2, \ldots, z_n]$, where $\mathbf{K}$ is any field. Then, $V(S) = V(I)$, which implies that every algebraic set is equal to $V(I)$ for some ideal I.

Definition 1.3. The polynomial **ideal** $I(Q)$ of a point set $Q \subset \mathbf{R}^n$ (or $Q \subset \mathbf{C}^n$) is the set of polynomials which vanish in Q. If f and g are polynomials in $I(Q)$, so is $f + g$ and so is ϕf for any polynomial ϕ. Any set S of polynomials defines a variety $V = V(S)$; S generates an ideal I consisting of linear combinations of elements of S with polynomial coefficients and $S \subset I(V(S)), Q \subset V(I(Q)), V(I(V(S))) = V(S), I(Q) = I(V(I(Q)))$.

The division algorithm, originating with Euclid, is crucial to many of the nice features of analysis and synthesis in 1-D system theory. A *principal ideal domain (PID)* is necessary for the division rule to be applicable. Unfortunately, $R[z_1, z_2, \ldots, z_n]$ is a PID if and only if R is a field and $n = 1$. A PID is a special case of a **Noetherian ring**, which is defined to be one whose every ideal is finitely generated. Much of the theory of polynomial rings is concerned with the question regarding how much of the nice properties of a Euclidean domain (which is, necessarily, a PID) can be recovered for non-Euclidean rings. One classical result is the **Hilbert Basis Theorem**, which states that if R is a Noetherian ring, then the multivariate polynomial ring $R[z_1, z_2, \ldots, z_n]$ is also Noetherian.

Next, consider the situation when either $\mathbf{K}$ is an algebraically closed field (and, therefore, this field is infinite) or $\mathbf{K}_1$ is an algebraically closed extension field of $\mathbf{K}$ (in the complementary situation). The existence of such a $\mathbf{K}_1$ for any field $\mathbf{K}$ is guaranteed by the axiom of choice. For our discussion here, we choose $\mathbf{K}_1 = \mathbf{C}$, the algebraically closed field of complex numbers. The set of zeros of any n-variate polynomial, in that case, is uncountably infinite. Furthermore, as a significant departure from the univariate polynomial case, the zero-set of a n-variate polynomial is unbounded and belongs to continuous algebraic curves instead of being composed of isolated zero points. In fact, no polynomial (or, for that matter, no holomorphic function) of more than one complex variable has any isolated zeros. One can prove, using an advanced result known as the *Weierstrass Preparation Theorem*, that the zero-set of a *non-constant* analytic function in C^n, the n-fold Cartesian product of C, is a complex hypersurface of dimension $(n - 1)$, except on a singular set of lower dimension. The *well-known* fact that the *zero-set* of a polynomial (or, for that matter, a holomorphic function) in one complex variable (that is, $n = 1$) is discrete (that is, a set of complex dimension 0) becomes, then, a valid specialization of this multivariate result.

Example 1.3. The zero-set of the polynomial,

$$a(z_1, z_2) = z_1^2 z_2^2,$$

is a one-dimensional complex surface (note that here $n = 2$), except at the point $(0,0)$ (which is zero-dimensional), of intersection of the two hyperplanes,

$$\{(z_1, z_2) \in C^2 : z_1 = 0\} \ \text{and} \ \{(z_1, z_2) \in C^2 : z_2 = 0\}.$$

Next, the reader is prepared for an important theorem due to Hilbert on zeroes of multivariate polynomials which supplies the precise relationship between algebraic sets and polynomial ideals. For this, it is convenient to have the following fundamental definitions from algebraic geometry. Most of the results in algebraic geometry are over the field of complex numbers though some results are also available over the field of real numbers [10].

Definition 1.4. Let R be a ring, I an ideal of R. Then the set

$$\{f \in R : f^r \in I \ for \ some \ nonnegative \ integer \ r\}$$

is called the **radical** of I and is denoted by $rad(I)$.

The **Hilbert Nullstellensatz** characterizes the radical of an ideal I of $K[z_1, z_2, \ldots, z_n]$, as the set of all polynomials that vanish at every zero of I in K_1^n, where K_1 is any specified algebraically closed extension of K.

Theorem 1.1 (Hilbert Nullstellensatz). *Let* K *be a field,* K_1 *an algebraically closed extension of* K, *and* $f, g_1, g_2, \ldots, g_r \in K[z_1, z_2, \ldots, z_n]$. *For all* $z \in K_1^n$, $g_1(z) = g_2(z) = \ldots = g_m(z) = 0$ *implies* $f(z) = 0$, *if and only if there exists a positive integer* r *such* f^r *belongs to the ideal generated by the n-variate polynomials* g_k, $k = 1, 2, \ldots, m$.

Interplay between geometric notions and algebraic concepts is provided by the correspondence between prime ideals and irreducible varieties (if I is a prime ideal, then V(I) is irreducible), radical ideals and affine algebraic varieties (if I is a radical ideal, the I(V(I)) = I), and maximal ideals and points. It should be noted that though the results of Hilbert just quoted are very powerful, they were of limited scope during the early years of multidimensional systems. In applications, constructive procedures are very often needed and the development of computational tools for algorithmic implementation of existential results in polynomial algebra has met with considerable success in recent years.

1.2.1 Multivariate Polynomial Factorization

Let D be an UFD and express $a(z_1, z_2, \ldots, z_n) \in D[z_1, z_2, \ldots, z_n]$ in recursive canonical form in the main variable, say z_n, of degree $d_{z_n} = m_n$, as

$$a(z_1, z_2, \ldots, z_n) = \sum_{k_n=0}^{m_n} a_{k_n}(z_1, z_2, \ldots, z_{n-1}) z_n^{k_n},$$

where $a_{k_n}(z_1, z_2, \ldots, z_{n-1}) \in D[z_1, z_2, \ldots, z_{n-1}]$ for $k_n = 0, 1, 2, \ldots m_n$. In this representation, the content of $a(z_1, z_2, \ldots, z_n)$ is the greatest common divisor (gcd) of the set of polynomial coefficients $\{a_{k_n}(z_1, z_2, \ldots, z_{n-1})\}$. The polynomial $a(z_1, z_2, \ldots, z_n) \in D[z_1, z_2, \ldots, z_{n-1}[z_n]]$ is primitive if it does not have a nontrivial factor in $D[z_1, z_2, \ldots, z_{n-1}]$.

Unique factorization is known to hold in the ring of integers, in the ring of polynomials in one or more variables with coefficients in a field, in the ring of p-adic integers, and in the ring of all formal power series in one or more variables, again with coefficients in a field. The conjecture suggested by the last two cases, that unique factorization holds in any local ring, was reduced to the case of dimension 3 by Nagata [11] and then was proved in this difficult case by Auslander and Buschbaum using techniques of homological algebra [12].

1.3 Multivariate Rational Functions

The field of quotients formed from any two elements (for which this quotient is defined) in the polynomial ring $\mathbf{K}[z_1, z_2, \ldots, z_n]$ over a field $\mathbf{K}$ (recall that this polynomial ring is a UFD) is denoted by $\mathbf{K}(z_1, z_2, \ldots, z_n)$ and is called the field of n-variate rational functions in the independent complex variables or indeterminates $z_1, z_2, \ldots, z_n$ over $\mathbf{K}$. Thus, like in the univariate case, a multivariate rational function is defined to be a quotient of a numerator polynomial and a *non-zero* denominator polynomial. A rational function is said to be in *reduced form* provided the numerator and denominator polynomials are coprime i.e. devoid of any nontrivial (degree zero) common factor. In the univariate case that is equivalent to the absence of any common zero in the two polynomials. However, in the multivariate case the *zero-set* of two coprime (relatively prime) polynomials may intersect. In that case the corresponding rational function is said to have a type of singularity referred to as *the nonessential singularity of the second kind*. The *zero-set* of the denominator polynomial of a rational function in reduced form that excludes the subset of zeros which constitute the nonessential singularities of the second kind comprise a type called *the nonessential singularity of the first kind*. Rational functions do not have any essential singularity.

Example 1.4. The bivariate rational function, $\frac{z_1}{(z_2-1)}$, is in reduced form and has a nonessential singularity of the second kind at location $(0, 1)$ in complex 2-*space*. The nonessential singularities of the first kind are uncountably infinite and occur at $(a, 1), a \neq 0$ and $(0, b), b \neq 1$.

The bivariate rational function, $\frac{1}{(z_1+z_2)}$ has a nonessential singularity of the second kind at (∞, ∞) but no such singularity in any compact domain.

The number of nonessential singularities of the second kind in a bivariate rational function is finite; that is not so in a rational function of three or more complex variables. For n-variate rational functions in n complex variables, the locus of this type of singularity is of real dimension $(2n - 4)$ in a space of real dimension $2n$; therefore, these singularities cannot disconnect the space. In the $n = 2$ case, therefore, these singularities occur as isolated points. Denote the *zero-set* of a function $f(\mathbf{z}) \in \mathbf{K}(z_1, z_2, \ldots, z_n)$ by $Z(f)$. The following theorem is useful.

Theorem 1.2. *Suppose that $n > 1$, $A(\mathbf{z})$ and $B(\mathbf{z})$ are coprime polynomials in the n complex variables $\mathbf{z} = (z_1, z_2, \ldots, z_n)$, $A(\mathbf{0}) = B(\mathbf{0}) = 0$, and Ω is a neighborhood of C^n. Then*

1. $Z(A) \bigcap \Omega$ *is not a subset of* $Z(B) \bigcap \Omega$;
2. *if* $\frac{f(\mathbf{z}) = A(\mathbf{z})}{B(\mathbf{z})}$ *and* $\gamma \in C$, *then there exists a point* $\mathbf{z} \in \Omega$ *such that* $B(\mathbf{z}) \neq 0$ *and* $f(\mathbf{z}) = \gamma$.

The proof for the above theorem is available in Rudin [13].

1.4   Relative Primeness

Definition 1.5. An element $b(z_1, z_2, \ldots, z_n)/a(z_1, z_2, \ldots, z_n)$ in $I(z_1, z_2, \ldots, z_n)$ (or $K(z_1, z_2, \ldots, z_n)$) will be said to be *of reduced form* if polynomials $b(z_1, z_2, \ldots, z_n)$, $a(z_1, z_2, \ldots, z_n)$ are relatively prime in $I[z_1, z_2, \ldots, z_n]$ (or $K[z_1, z_2, \ldots, z_n]$), i.e., are devoid of common polynomial factors other than units.

1.4.1   Tests for Relative Primeness

To test for n-variate polynomial relative primeness, using the classical results based on the theory of resultants, the numerator and denominator polynomials are each written in recursive canonical form in the main variable, say z_1:

$$b(z_1, z_2, \ldots, z_n) = \sum_{k=0}^{r_1} b_k(z_2, z_3, \ldots, z_n) z_1^k$$

$$a(z_1, z_2, \ldots, z_n) = \sum_{k=0}^{m_1} a_k(z_2, z_3, \ldots, z_n) z_1^k. \tag{1.1}$$

Without any loss of generality, it may be assumed that the polynomial sequence $\{b_0, \ldots, b_{r_1}, a_0, \ldots, a_{m_1}\}$ is devoid of a common factor. It is also assumed that b_k's and a_k's belong to a unique factorization domain $D[z_2, z_3, \ldots, z_n]$, and that $b_{r_1}(z_2, z_3, \ldots, z_n) \neq 0$ and $a_{m_1}(z_2, z_3, \ldots, z_n) \neq 0$, $r_1 > 0$, $m_1 > 0$. Form the matrix $\mathbf{R}$ of order $r_1 + m_1$ involving b_k's and a_k's, the indeterminates not being shown for brevity, as follows.

$$\mathbf{R} = \begin{bmatrix} b_{r_1} & b_{r_1-1} & \cdots & b_1 & b_0 & 0 & 0 & \cdots & 0 \\ 0 & b_{r_1} & b_{r_1-1} & \cdots & b_1 & b_0 & 0 & \cdots & 0 \\ 0 & 0 & b_{r_1} & b_{r_1-1} & \cdots & b_1 & b_0 & \cdots & 0 \\ & & & \cdots\cdots\cdots\cdots & & & & & \\ 0 & 0 & \cdots & 0 & b_{r_1} & b_{r_1-1} & \cdots & b_1 & b_0 \\ 0 & \cdots & 0 & a_{m_1} & a_{m_1-1} & \cdots & \cdots & a_1 & a_0 \\ & & & \cdots\cdots\cdots\cdots & & & & & \\ 0 & a_{m_1} & a_{m_1-1} & \cdots & \cdots & a_1 & a_0 & \cdots & 0 \\ a_{m_1} & a_{m_1-1} & \cdots & \cdots & a_1 & a_0 & 0 & \cdots & 0 \end{bmatrix} \tag{1.2}$$

$\mathbf{R}$ is referred to as a *Sylvester* or *inner matrix*.

Theorem 1.3. *With $a_{m_1} \neq 0$, $b_{r_1} \neq 0$, $r_1 > 0$, $m_1 > 0$, $b(z_1, \ldots, z_n)$ and $a(z_1, \ldots, z_n)$ in (1.1) are relatively prime if and only if the resultant $\det \mathbf{R} \triangleq r(z_2, z_3, \ldots, z_n) \neq 0$. It is assumed that the sequence $\{b_0, \ldots, b_{r_1}, a_0, \ldots, a_{m_1}\}$ is devoid of a common factor.*

Proof. Observe that because the sequence $\{b_0, \ldots, b_{r_1}, a_0, \ldots, a_{m_1}\}$ is devoid of a common factor, any factor common to the two given polynomials must involve the variable z_1. Consequently, the given polynomials in (1.1) have a nonconstant common factor if and only if there exist two polynomials $f(z_1, \ldots, z_n)$ and $g(z_1, \ldots, z_n)$ such that

$$f(z_1, \ldots, z_n) a(z_1, \ldots, z_n) = g(z_1, \ldots, z_n) b(z_1, \ldots, z_n) \tag{1.3}$$

with

$$\delta_{z_1}[f(z_1, \ldots, z_n)] < \delta_{z_1}[b(z_1, \ldots, z_n)] \tag{1.4}$$

$$\delta_{z_1}[g(z_1, \ldots, z_n)] < \delta_{z_1}[a(z_1, \ldots, z_n)].$$

After writing all polynomials in recursive canonical form in the main variable z_1, (1.3) leads to a system of equations which has a nontrivial solution for $f(z_1, \ldots, z_n)$, $g(z_1, \ldots, z_n)$ if and only if $\det \mathbf{R} \equiv 0$. Therefore $a(z_1, \ldots, z_n)$, $b(z_1, \ldots, z_n)$ are relatively prime if and only if $\det \mathbf{R} \neq 0$.

The above result is a generalization of the result for the $n = 1$ case [14, pp. 83–85]. Note that $\det \mathbf{R} \equiv 0$ also when $a_{m_1} \equiv 0$ and $b_{r_1} \equiv 0$. The test for absence of common factor in $\{b_0, \ldots, b_{r_1}, a_0, \ldots, a_{m_1}\}$ is a special of the result just given in the sense that a lesser number of indeterminates are involved.

Other tests for relative primeness are also available [6]. Some of these tests, along with schemes for g.c.f. extraction when polynomials are not relatively prime, will be discussed later in this chapter.

1.4.2 Primitive Factorization Algorithms for g.c.f. Extraction

Without any loss of generality, it can be assumed that the set of b_k's and the set of a_k's in (1.1) each have no common factors. In case this assumption of primitivity with respect to main variable z_1 in $a(z_1, z_2, \ldots, z_n)$ and $b(z_1, z_2, \ldots, z_n)$ is not satisfied, the content of each can be found by applying repeatedly the test to be discussed here on the $(n-1)$-variate polynomials $(a_0, a_1, \ldots, a_{m_1})$ when finding the content of $a(z_1, \ldots, z_n)$. It is well known that

$$\text{g.c.f.}(a, b) = \{\text{g.c.f.}[\text{cont}(a), \text{cont}(b)]\} \, \{\text{g.c.f.}[\text{pp}(a), \text{pp}(b)]\}, \tag{1.5}$$

where pp[·] denotes the "primitive part of [·]." Therefore, unless stated otherwise the given polynomials in (1.1) will be assumed to be primitive.

1.4.2.1 Multivariate g.c.d. Extraction from the Sylvester Matrix

Running parallel to the established results in the single-variable case, the innerwise [15] matrix $[\triangle (a, b)_{i,j}]$ associated with (1.1) is written as follows:

$$[\triangle (a, b)_{i,j}] \triangleq \begin{bmatrix} a_{m_1} & a_{m_1-1} & a_{m_1-2} & \cdots & a_{m_1+1-i-j} \\ 0 & a_{m_1} & a_{m_1-1} & \cdots & a_{m_1+2-i-j} \\ 0 & 0 & a_{m_1} & \cdots & a_{m_1+3-i-j} \\ & & \cdots\cdots\cdots\cdots & & \\ 0 & 0 & b_{r_1} & \cdots & b_{3+r_1-i-j} \\ 0 & b_{r_1} & b_{r_1-1} & \cdots & b_{2+r_1-i-j} \\ b_{r_1} & b_{r_1-1} & b_{r_1-2} & \cdots & b_{1+r_1-i-j} \end{bmatrix}. \tag{1.6}$$

In (1.6), there are i rows containing the a_k's, j rows containing the b_k's, and $a_i \equiv 0$, $i < 0$, $b_j \equiv 0$, $j < 0$. Next, consider the following n-variate subresultant:

$$\Delta(a,b,z_1)_{i,j} \equiv \det \begin{bmatrix} a_{m_1} & a_{m_1-1} & \cdots & a_{m_1+2-i-j} & z_1^{i-1}a \\ 0 & a_{m_1} & \cdots & a_{m_1+3-i-j} & z_1^{i-2}a \\ & & \cdots\cdots\cdots\cdots & & \\ 0 & b_{r_1} & \cdots & b_{3+r_1-i-j} & z_1^{j-2}b \\ b_{r_1} & b_{r_1-1} & \cdots & b_{2+r_1-i-j} & z_1^{j-1}b \end{bmatrix} \qquad (1.7)$$

obtained by replacing only the last column in (1.6) by the last column defined in (1.7). The new column can be expanded as follows:

$$\Delta(a,b,z_1)_{i,j} = f(z_1,z_2,\ldots,z_n)a(z_1,z_2,\ldots,z_n)$$
$$+ g(z_1,z_2,\ldots,z_n)b(z_1,z_2,\ldots,z_n) \qquad (1.8)$$

where the degrees in z_1 of $f(z_1,\ldots,z_n)$ and $g(z_1,\ldots,z_n)$ are, respectively $i-1$ and $j-1$. Using the fact that the g.c.f. of two primitive polynomials is primitive, the result stated next follows. Note that $\det[\Delta(a,b)_{r_1,m_1}]$ is the resultant of the two polynomials in (1.1).

Theorem 1.4. *Suppose for two primitive n-variate polynomials, $a(z_1,z_2,\ldots,z_n)$ and $b(z_1,z_2,\ldots,z_n)$ in (1.1), the associated $[\Delta(a,b)_{i,j}]$ and $\Delta(a,b,z_1)_{i,j}$ are as defined in (1.6), (1.7). Also, define $\Delta(a,b)_{i,j} \equiv \det[\Delta(a,b)_{i,j}]$. Then if,*

$$\Delta(a,b)_{r_1,m_1} \equiv \Delta(a,b)_{r_1-1,m_1-1} \equiv \cdots \equiv \Delta(a,b)_{r_1-j+1,m_1-j+1} \equiv 0 \qquad (1.9)$$

and

$$\Delta(a,b)_{r_1-j,m_1-j} \neq 0, \qquad (1.10)$$

then g.c.f. (a,b) is of degree j in z_1 and,

$$\text{g.c.f.}(a,b) \equiv \text{pp}\{\Delta(a,b,z_1)_{r_1-j,m_1-j}\}. \qquad (1.11)$$

Conversely, if g.c.f(a,b) is of degree j in z_1 then it is given by (1.11), and (1.9) and (1.10) are valid.

The proof of the above theorem can be constructed from the arguments given a priori.

Example 1.5. It is required to extract the g.c.f. from

$$a(z_1,z_2,z_3) = z_2^2 z_1^4 + (z_2^3 + 2z_2^2 z_3)z_1^3 + (1 + z_2^3 z_3 + z_3^2 z_2^2)z_1^2$$
$$+ (z_2 + 2z_3)z_1 + (z_2 z_3 + z_3^2)$$
$$b(z_1,z_2,z_3) = (z_2 + z_3^4)z_1^2 + (2z_3^5 + z_2 z_3^4 + 2z_2 z_3 + z_2^2)z_1$$
$$+ (z_3^6 + z_2 z_3^5 + z_2 z_3^2 + z_2^2 z_3),$$

if they are not relatively prime.

Considering z_1 as the main variable,

$$\text{cont}\{a(z_1, z_2, z_3)\} = \text{g.c.f.}\{a_4, a_3, a_2, a_1, a_0\}$$

$$\text{cont}\{b(z_1, z_2, z_3)\} = \text{g.c.f.}\{b_2, b_1, b_0\}$$

The fact that $\text{cont}\{a(z_1, z_2, z_3)\} = 1$ is verifiable by inspection, since $a_4 = z_2 \cdot z_2$, $a_0 = z_3(z_2 + z_3)$ are relatively prime. Also, $\text{cont}\{b(z_1, z_2, z_3)\} = \text{g.c.f.}\{b_2, b_1, b_0\} = (z_2 + z_3^4)$. Therefore,

$$\text{g.c.f.}[\text{cont}(a), \text{cont}(b)] = 1. \tag{1.12}$$

With respect to the main variable z_1

$$\text{pp}\{a(z_1, z_2, z_3)\} = a(z_1, z_2, z_3)$$

$$\text{pp}\{b(z_1, z_2, z_3)\} = z_1^2 + (z_2 + 2z_3)z_1 + (z_2 z_3 + z_3^2)$$

For $\text{pp}\{a(z_1, z_2, z_3)\}$ and $\text{pp}\{b(z_1, z_2, z_3)\}$, the matrix corresponding to (1.6) is formed. It is found that $\triangle[\text{pp}(a), \text{pp}(b)]_{2,4} \equiv 0$, implying that $\text{pp}(a)$ and $\text{pp}(b)$ are not relatively prime. Also,

$$\triangle[\text{pp}(a), \text{pp}(b)]_{1,3} \equiv 0$$

$$\triangle[\text{pp}(a), \text{pp}(b)]_{0,2} = 1 \neq 0,$$

implying that the degree in z_1 of the g.c.f. of $\text{pp}(a)$ and $\text{pp}(b)$ is 2. The g.c.f. of $\text{pp}(a)$ and $\text{pp}(b)$ is

$$\begin{aligned}
\text{g.c.f.}[\text{pp}(a), \text{pp}(b)] &= \text{pp} \det \begin{bmatrix} 0 & \text{pp}\{b\} \\ 1 & z_1 \text{pp}\{b\} \end{bmatrix} \\
&= z_1^2 + (z_2 + 2z_3)z_1 + (z_2 z_3 + z_3^2) \\
&= (z_1 + z_2 + z_3)(z_1 + z_3).
\end{aligned} \tag{1.13}$$

Using (1.5), it follows from (1.12) and (1.13) that

$$\text{g.c.f.}[a(z_1, z_2, z_3), b(z_1, z_2, z_3)] = (z_1 + z_2 + z_3)(z_1 + z_3)$$

1.4.2.2 Multivariate g.c.d. Extraction from the Bezout Matrix

Remember that, without loss of generality, $a(z_1, \mathbf{z})$ and $b(z_1, \mathbf{z})$ are primitive in $K[\mathbf{z}][z_1]$, and a not too interesting trivial case is avoided by requiring $a_0(\mathbf{z})$ and $b_0(\mathbf{z})$ to be not zero polynomials. For the sake of brevity, $a_k(\mathbf{z})$, $b_k(\mathbf{z})$ will be denoted simply by a_k and b_l and let

$$|a_k b_l| = a_k b_l - a_l b_k. \tag{1.14}$$

Associate the Bezout matrix $B(a, b)$ below with the polynomials in (1.1). $B(a, b)$ is symmetric and square of order m_1, where without loss of generality, $r_1 \leq m_1$.

$$B(a, b) = \begin{bmatrix} |a_0b_1| & |a_0b_2| & |a_0b_3| & \cdots \\ |a_0b_2| & |a_0b_3| + |a_1b_2| & |a_0b_4| + |a_1b_3| & \cdots \\ |a_0b_3| & |a_0b_4| + |a_1b_3| & |a_0b_5| + |a_1b_4| + |a_2b_3| & \cdots \\ \cdots & \cdots & \cdots & \cdots \end{bmatrix} \qquad (1.15)$$

Fact 1.1. *The nullity of $B(a, b)$ is equal to the partial degree in the main variable z_1 of the greatest common divisor of $a(z_1, z_2)$ and $b(z_1, z_2)$.*

Suppose that the rank of $B(a, b)$ is r, so that its nullity is $(m_1 - r)$. Next, define the matrix $C(a, b)$ of order r in (1.16) below. For the sake of brevity, the matrix in (1.16) has been written to correspond to the case when $r = 3$.

$$C(a, b) = \begin{bmatrix} |a_0b_1| & |a_0b_2| & |a_0b_3| + |a_0b_4|z_1 \\ |a_0b_2| & |a_0b_3| + |a_1b_2| & |a_0b_4| + |a_1b_3| + (|a_0b_5| + |a_1b_4|)z_1 \\ |a_0b_3| & |a_0b_4| + |a_1b_3| & |a_0b_5| + |a_1b_4| + |a_2b_3| + (|a_0b_6| + |a_1b_5| + |a_2b_4|)z_1 \end{bmatrix}$$

$$(1.16)$$

Fact 1.2. *The greatest common divisor of the primitive multivariate polynomials $a(z_1, \mathbf{z})$ and $b(z_1, \mathbf{z})$ is given by the primitive part in $K[\mathbf{z}][z_1]$ of the determinant of matrix $C(a, b)$.*

The proof may be constructed by exploiting the relationship between a Sylvester matrix and a Bezout matrix. The two are linked through a nonsingular linear transformation matrix whose elements belong to $K[\mathbf{z}]$.

Example 1.6. Consider bivariate polynomials $a(z_1, z_2)$ and $b(z_1, z_2)$ shown below to be elements of $\mathcal{R}[z_2][z_1]$.

$$a(z_1, z_2) = z_2 + (z_2 + 1)z_1 + z_1^2$$
$$b(z_1, z_2) = 1 + (z_2 + 1)z_1 + z_2z_1^2$$

Clearly, $m_1 = r_1 = 2$ and

$$a_0(z_2) = z_2, \quad a_1(z_2) = z_2 + 1, \quad a_2(z_2) = 1$$
$$b_0(z_2) = 1, \quad b_1(z_2) = z_2 + 1, \quad b_2(z_2) = z_2.$$

Clearly, both are primitive in $\mathcal{R}[z_2][z_1]$.

The Bezout matrix $B(a, b)$ for this case is,

$$B(a, b) = \begin{bmatrix} z_2^2 - 1 & z_2^2 - 1 \\ z_2^n & z_2^2 - 1 \end{bmatrix}$$

The nullity of $B(a, b)$ is 1 and its rank is 1. The partial degree in z_1 of the g.c.d., therefore, is 1. The g.c.d. can be computed to be

$$\text{pp}[|a_0 b_1| + |a_0 b_2| z_1] = \text{pp}[(z_2^n) + (z_2^2 - 1) z_1]$$

$$= (z_1 + 1)$$

Example 1.7. It is required to find the common zeros of the two polynomials below. The polynomials are written in $\mathbb{R}[\mathbf{x_2}][\mathbf{x_1}]$.

$$f(x_1, x_2) = (x_2 + 1) x_1^2 + (2x_2) x_1 + x_2^3 \tag{1.17}$$

$$g(x_1, x_2) = x_1^2 - 6x_1 - 3x_2^2 \tag{1.18}$$

The resultant matrix for the two polynomials is:

$$\mathbf{R}(x_2) = \begin{pmatrix} (x_2 + 1) & 2x_2 & x_2^3 & 0 \\ 0 & (x_2 + 1) & 2x_2 & x_2^3 \\ 0 & 1 & -6 & -3x_2^2 \\ 1 & -6 & -3x_2^2 & 0 \end{pmatrix}$$

Then the resultant $det\ \mathbf{R}(x_2) = -x_2^4 (4x_2 + 3)^2$ has zeros at $x_{21} = 0$, $x_{22} = -\frac{3}{4}$.

Consider, first, $x_{21} = 0$. Then, $f(x_1, 0) = x_1^2$, $g(x_1, 0) = x_1^2 - 6x_1$ and gcd $[f(x_1, 0), g(x_1, 0)] = x_1$. Therefore there is a common zero at $(0,0)$.

For $x_{22} = -\frac{3}{4}$, $g(x_1, -\frac{3}{4}) = x_1^2 - 6x_1 - \frac{27}{16}$, $f(x_1, -\frac{3}{4}) = \frac{x_1^2}{4} - \frac{3}{2}x_1 - \frac{27}{64}$. gcd $[f(x_1, -\frac{3}{4}), g(x_1, -\frac{3}{4})] = x_1^2 - 6x_1 - \frac{27}{16}$ has zeros at $x_{10} = 3 + \frac{3}{4}\sqrt{19}$ and $x_{20} = 3 - \frac{3}{4}\sqrt{19}$. Therefore, the common zeros are at $(3 + \frac{3}{4}\sqrt{19}, -\frac{3}{4})$ and $(3 - \frac{3}{4}\sqrt{19}, -\frac{3}{4})$.

Note that the subresultant obtained by calculating the determinant of the (2×2) submatrix obtained after deletion of the bordering columns and rows of $\mathbf{R}$ is $(-8x_2 - 6)$, which is zero when $x_2 = -\frac{3}{4} = x_{22}$. This justifies the degree of $gcd[f(x_1, -\frac{3}{4}), g(x_1, -\frac{3}{4})]$.

1.5   Holomorphic Functions in Several Complex Variables

Complex analysis in several complex variables is done on C^n, the n-dimensional vector space over the complex space C. Topologically, C^n is the real Euclidean space R^{2n}. The terms analytic and holomorphic are synonymous for *complex-valued* functions of complex variables. In function theory of real or complex variables, the descriptor *analytic* is associated with the feasibility for power series representation of a function while *holomorphic* signifies differentiability. We summarize here results which can be viewed as generalizations of their univariate counterparts as well as several distinguishing features in the analyticity (holomorphicity) theory of several complex variables ($n > 1$) that complicate matters in the use of such theory in applications of concern to us. Typical examples of holomorphic functions are polynomials and rational functions considered in the previous two sections.

A fundamental and deep theorem due to Hartogs unifies various definitions for holomorphic functions in several complex variables into an equivalent one.

Theorem 1.5 (Hartogs). *Let $\Omega \subset \mathbf{C}^n$ be an open set and suppose that f is a function in n complex variables that maps Ω into $\mathbf{C}$. Suppose that for each arbitrary but fixed set of the $(n-1)$ complex variables, $z_1, \ldots, z_{i-1}, z_{i+1}, \ldots z_n$, $i = 1 \ldots n$, the function, $f(\bullet) = f(z_1, \ldots, z_{i-1}, \bullet, z_{i+1}, \ldots z_n)$, is an univariate holomorphic function. Then the function f is continuous on Ω.*

The appropriate counterpart of the above theorem does not hold in the case of analytic functions of several real variables.

Definition 1.6. A *complex-valued* function $f(\mathbf{z})$ defined in some open set $\Omega \subset \mathbf{C}^n$ is said to be holomorphic (analytic) in Ω if and only if $f(\mathbf{z})$ is analytic in each variable separately.

A convergent power series in several complex variables is another example of an holomorphic function for which the determination of the largest domain of holomorphy is, in general, a hard problem [16].

Definition 1.7. The convergence domain of a power series is the set of points like $\mathbf{z}^0$ such that the power series is absolutely convergent in a neighbourhood of $\mathbf{z}^0$.

Unlike the univariate case, where the domain of convergence of a power series is always an unit disc, there is no such simple geometrical configuration for the domain of convergence of a power series in several complex variables.

Definition 1.8. The set of points in $\mathbf{C}^n$ where a n-variate power series is absolutely convergent constitutes the **Reinhardt domain** or the region of analyticity of the function represented by the power series.

The Reinhardt domain is, in general, greater than the convergence domain and depends only on the magnitudes of the n independent complex variables (which implies that the domain is in $\mathbf{R}^n$). The Reinhardt domain is also *logarithmically convex* in the sense that it is mapped onto a convex domain in $\mathbf{R}^n$ by the mapping $z_i \longrightarrow \log z_i$, for $i = 1, 2, \ldots, n$.

According to the Riemann Mapping Theorem, every proper simply connected open subset $\Omega \subset \mathbf{C}$ is biholomorphic to the unit disc in the sense that there exists a holomorphic function that maps Ω bijectively (one-to-one and onto) to the unit disc. There is no such canonical domain like the unit disc in $\mathbf{C}^n$, when $n \geq 2$. Therefore, analysis of holomorphic functions of several variables depends on the domain in question. Two extensively studied domains in $\mathbf{C}^n$ are the ball,

$$B(\mathbf{z_0}, r) = \{\mathbf{z} \in C^n : |\mathbf{z} - \mathbf{z_0}| < \mathbf{r}\},$$

and the polydisc,

$$D^n(\mathbf{z}^0, \ r) \ = \ \{\mathbf{z} \in C^n \ : |z_i - z_i^0| < r\} , \ i = 1, \ldots, n.$$

Let U denote the unit disc $D^1(0, \ 1)$ and let $\bar{U}$ denote the closure of U. $\bar{U}^n$ denotes the closure of the polydisc

$$D^n(\mathbf{0}, \ 1) \ \triangleq \ \mathbf{U^n} = \mathbf{U} \times \mathbf{U} \ldots \times \mathbf{U}.$$

The distinguished boundary of the polydisc is

$$T^n \ \triangleq \ \{\mathbf{z} \in \mathbf{C^n} \ : \ |\mathbf{z_1}| = |\mathbf{z_2}| \ = \ldots = \ |\mathbf{z_n}| = 1\}.$$

The polydisc algebra is fundamental in the analysis and design of a class of multidimensional discrete-space systems characterized by the linear and shift-invariance properties.

A consequence of Definition 1.6 is that a holomorphic function in $\Omega \subset \mathbf{C}^n$ is necessarily continuous in Ω and also continuously differentiable in the $2n$ real variables, x_k, y_k, where

$$z_k = x_k + y_k \, for \, k = 1, 2, \ldots, n.$$

An analytic function can be expanded as a power series in the neighbourhood of a point where it is well-defined by calculating the value of the function and its derivatives at that point. Furthermore, it is easy to see that every univariate analytic function is equivalent to a polynomial in the neighbourhood of a critical point. Norman Levinson proved an analogous result for functions of two variables [17]. Interestingly, Hassler Whitney has given examples of analytic functions of three variables that are not equivalent to a polynomial in the neighbourhood of a critical point [18]. For one such example, consider the trivariate function,

$$a(z_1, z_2, z_3) = z_1 z_2 (z_1 - z_2)(z_1 - z_2 z_3)(z_1 - z_2 \exp z_3).$$

It has further been shown [19] that a function $f(z_1, \ldots, z_{n-1}, z_n)$ may be transformed into a polynomial in z_n with coefficients that are analytic functions of the other indeterminates. However, it cannot in general (say with $n = 2$) be transformed into a polynomial.

Let the real and imaginary parts of the complex variable z_i be x_i and y_i; i. e. $z_i = x_i + j \, y_i$. Define two operators below.

$$\frac{\delta}{\delta z_i} = \frac{1}{2} \left(\frac{\delta}{\delta x_i} - j \frac{\delta}{\delta y_i} \right)$$

and

$$\frac{\delta}{\delta \bar{z}_i} = \frac{1}{2} \left(\frac{\delta}{\delta x_i} + j \frac{\delta}{\delta y_i} \right).$$

Then, the preceding definition leads to the following theorem, which is the counterpart of the *Cauchy-Riemann* conditions in the univariate case.

Theorem 1.6. *A continuously differentiable function $f(\mathbf{z})$ in n complex variables denoted by $\mathbf{z}$ is holomorphic if and only if*

$$\frac{\delta}{\delta \bar{z}_i} = 0, \quad i = 1, \ldots, n.$$

Let $u(\mathbf{z}, \bar{\mathbf{z}})$ and $v(\mathbf{z}, \bar{\mathbf{z}})$ be the real and imaginary parts of a *complex-valued* n-variate holomorphic function $f(\mathbf{z})$. Then, for $i, k = 1, 2, \ldots, n$, the **pluriharmonic** conditions on the functions $u(\mathbf{z}, \bar{\mathbf{z}})$ and $v(\mathbf{z}, \bar{\mathbf{z}})$ are, respectively, given below.

$$\frac{\delta^2 u(\mathbf{z}, \bar{\mathbf{z}})}{\delta z_i \delta \bar{z}_k} = 0, \tag{1.19}$$

and

$$\frac{\delta^2 v(\mathbf{z}, \bar{\mathbf{z}})}{\delta z_i \delta \bar{z}_k} = 0, \tag{1.20}$$

Recalling that $z_i = x_i + j\, y_i, i = 1, 2, \ldots, n$, introduce the operator,

$$\Delta_i = \frac{\delta^2}{\delta x_i^2} + \frac{\delta^2}{\delta y_i^2} \tag{1.21}$$

An harmonic function $f(\mathbf{z})$ is one for which

$$\sum_{i=0}^{n} \Delta_i = 0. \tag{1.22}$$

A n-harmonic function is defined next.

Definition 1.9. A continuous complex function $f(\mathbf{z})$ in an open subset of $\mathbf{C}^n$ is called n-harmonic if it is harmonic in each of the complex variables separately i. e. $\Delta_i f = 0$, for $i = 1, 2, \ldots, n$.

Clearly, the classes of harmonic and n-harmonic functions coincide if and only if $n = 1$.

A univariate holomorphic function $f(z)$ in any given domain in the complex plane is not continuable to the exterior of the domain if all the boundary points of the domain are singularities of $f(z)$. The boundary then is called the *natural boundary* of $f(z)$. The counterpart of the univariate result that for any open set in the complex plane there exists an analytic function whose natural boundary is the closure of this open set (and, therefore, the function cannot be analytically continued to a larger open set) does not hold in the several variable case. For an elementary discussion of this complex phenomenon and proofs of some of the consequences relevant in our interest, quoted below, see [16].

Fact 1.3. *If $f(\mathbf{z})$ is a holomorphic function in n complex variables $\mathbf{z}$ on a sub-domain of C^n, $n > 2$, then $f(\mathbf{z})$ has no isolated zeros.*

Fact 1.4. *Let $f(\mathbf{z})$ be a holomorphic function on a bounded sub-domain of $\mathbf{C}^n$, $n > 2$. Then the zero-set of $f(\mathbf{z})$ is either empty or non-compact there.*

We have already discussed the consequences of the preceding two facts in the special case of multivariate polynomials. There are some results in univariate complex analysis, like the maximum-modulus principle that generalize naturally to the several complex variable case.

Fact 1.5. *If $\Omega \subset \mathbf{C}^n$ is an open smoothly bounded domain and $f(\mathbf{z})$ is a non-constant function which is continuously differentiable on the closure $\bar{\Omega}$ of Ω and holomorphic on Ω, then the maximum modulus of $f(\mathbf{z})$ occurs on the boundary of Ω.*

In the univariate case, the image of a function is not always contained in the image of that function evaluated on the boundary of Ω. The situation is just the opposite in dimensions two and higher. With regard to other properties, subtle differences could occur when proceeding from 1-D to 2-D, and also from 2-D to 3-D. Consider, for example, the Hadamard product,

$$C(z) = \sum_{k=0}^{\infty} a_k b_k z^k ,$$

of two power series,

$$A(z) = \sum_{k=0}^{\infty} a_k z^k , \; B(z) = \sum_{k=0}^{\infty} b_k z^k$$

in one complex variable. Then, under certain restrictions [20], the singularities of $C(z)$ can be characterized with sufficient completeness from those of $A(z)$ and $B(z)$ i. e. the sets of singularities $\mathbf{A, B, C}$ of $A(z)$, $B(z)$, and $C(z)$, respectively, satisfy $\mathbf{C} \subset \mathbf{A} \mathbf{B}$. This is particularly true for the class of univariate rational functions, in which case the Hadamard product of two power series each having a rational representation also has a rational function representation. Significant differences occur in the bivariate and trivariate cases. In the bivariate case, the Hadamard product of two power series, each of which has a rational function representation, is known to have, in general, not a rational but an algebraic function representation while there are fundamental differences between the properties of the Hadamard product of rational functions of two and three variables [21].

1.6 Gröbner Bases

The topic of this section is concerned with a certain type of basis for polynomial ideals which provides an algorithmic method for solving a number of computability and decidability problems concerning the ideal; for example, given a multivariate polynomial $f(\mathbf{z})$ and an ideal I specified by a finite number of generators, one can perform computations in the original polynomial coefficient field to decide constructively whether or not $f(\mathbf{z})$ belongs to I. This author's interest in Gröbner bases originated during the nineteen eighties when Professor Buchberger, a former student of Gröbner, visited the University of Pittsburgh as an invitee of the author. On realizing the relevance of Gröbner bases to multidimensional systems theory, this author's invitation to contribute a survey chapter on the subject in a book [6] was gladly accepted by Bruno Buchberger. The highly readable Chap. 4 in [6] is strongly recommended to the readers of this text. Another worthwhile reading is a recent text [22], where computational algorithms for the fast construction of such bases are extensively documented. It is worth mentioning as a historical fact that the algorithm developed by Buchberger for the solution of a problem on the multivariate analog of the Euclidean algorithm posed by Gröbner led to the naming of Gröbner bases by the humble student in deference to his respected teacher. Justifiably, subsequent researchers have, sometimes, referred to Gröbner bases as *Gröbner–Buchberger bases*. Many computational problems that are extremely difficult for polynomial ideals generated by arbitrary bases are very easy for polynomial ideals generated by Gröbner bases. One instance of particular interest in this text is a formula expressing the greatest common divisor of a set of polynomials of several variables in terms of a Gröbner basis of the ideal generated by them [23].

Let a polynomial ring involving indeterminates $z_1, z_2, \ldots, z_n$ over an arbitrary but fixed base field $\mathbf{K}$ of coefficients be denoted by

$$\mathbf{R} = \mathbf{K}[z_1, z_2, \ldots, z_n].$$

A monomial $z_1^{i_1} z_2^{i_2} \ldots z_n^{i_n}$ will be called a *term* whose degree is $i_1 + i_2 + \ldots i_n$. Define an order $<_T$ on the terms as follows: $x_1^{i_1} \cdots x_n^{i_n} <_T x_1^{j_1} \cdots x_n^{j_n}$ if and only if

$deg\ (x_1^{i_1} \cdots x_n^{i_n})\ <\ deg\ (x_1^{j_1} \cdots x_n^{j_n})$ or $deg\ (x_1^{i_1} \cdots x_n^{i_n})\ =\ deg\ (x_1^{j_1} \cdots x_n^{j_n})$ and $i_1 = j_1, \ldots, i_k = j_k,\ i_{k+1} > j_{k+1}$ for some k with $1 \leq k + 1 \leq n$. The next example illustrates an order $<_T$ defined on the terms.

Example 1.8. $1 < z_1 < z_2 < z_3 < z_1^2 < z_1 z_2 < z_1 z_3 < z_2^2 < z_2 z_3 < z_3^2$
$< z_1^3 < z_1^2 z_2 < z_1^2 z_3 < z_1 z_2^2 < z_1 z_2 z_3 < z_1 z_3^2 < z_2^3$
$< z_2^2 z_3 < z_2 z_3^2 < z_3^3 < z_1^4 < \ \ldots.$

The above shows an admissible linear ordering of monomials. This could be lexicographical, graded lexicographical etc.

1.6.1　Notations

In the following notations t denotes a term, taken to be a monomial $z_1^{i_1} z_2^{i_2} \cdots z_n^{i_n}$ of degree $i_1 + i_2 + \cdots + i_n$ and f, g, are polynomials in $K[\mathbf{z}]$. $Coef(t,f)$ denotes the coefficient (possibly zero) of the term t in f. $Occ(t,f)$, or t occurs in f, if and only if $Coef(t,f) \neq 0$. These notations and others to be used here are listed next.

$Coef(t,f) \triangleq$ coefficient of the term t in polynomial f (possibly 0).
$Occ(t,f) \longleftrightarrow Coef(t,f) \neq 0.$
$Hcoef(f) \triangleq Coef(Hterm(f),f).$
$Head(f) \triangleq Hcoef(f)Hterm(f).$
$Rest(f) \triangleq f - Head(f).$
$Mult(z_1^{i_1} z_2^{i_2} \cdots z_n^{i_n}, z_1^{j_1} z_2^{j_2} \cdots z_n^{j_n}), \longleftrightarrow i_1 \geq j_1, i_2 \geq j_2 \ldots, i_n \geq j_n.$
$LCM(z_1^{i_1} z_2^{i_2} \cdots z_n^{i_n}, z_1^{j_1} z_2^{j_2} \cdots z_n^{j_n}), \triangleq z_1^{max(i_1,j_1)} z_2^{max(i_2,j_2)} \cdots z_1^{max(i_n,j_n)}.$

In the following notations F denotes a finite sequence $\{f_1, f_2, \ldots, f_r\}$ of polynomials.

$L(F) \triangleq$ length of the sequence $\{f_1, f_2, \ldots, f_r\}$.
$Mterm(t, F) \longleftrightarrow \exists f_i \in F, f_i \neq 0,$ such that $Mult(t, Hterm(f_i)).$
$Normalf(g, F) \longleftrightarrow \forall t$ with $Occ(t, g), \neg Mterm(t, F).$

Let $F = \{f_1, f_2\} = \{z_1^3 z_2 z_3 - z_1 z_2, z_1 z_2 z_3^2 - 2z_3^2\}$. Then,

$Head(f_1) = z_1^3 z_2 z_3, \quad Head(f_2) = z_1 z_2 z_3^2,$
$LCM(Head(f_1), Head(f_2)) = z_1^3 z_2 z_3^2,$
$Rest(f_1) = -z_1 z_2, \quad Rest(f_2) = -2z_3^2.$

Next, a reduction procedure is defined following the introduction of the following notations.

$f \bullet >_{F,t,i}^{(1)} g \longleftrightarrow 1 \leq i \leq L(F), f_i \neq 0, Occ(t,f), Mult(t, Hterm(f_i)), g = f - a \cdot s \cdot f_i,$ where $a = \frac{Coef(t,f)}{Hcoef(f_i)}, \ s = \frac{t}{Hterm(f_i)}.$

$f \bullet >_F^{(1)} g \longleftrightarrow \exists t, i$ such that $f \bullet >_{F,t,i}^{(1)} g.$

$f \bullet >_F g \longleftrightarrow \exists f_0, \ldots, f_k \in K[\mathbf{z}, k \geq 0,$ such that $f_0 = f, f_k = g, f_i \bullet >_F^{(1)} f_{i+1}, i = 0, \ldots, (k-1).$ (Notice that $f = g$ is included here. If $f \bullet >_F g$, then f M-reduces to g).

$f \bullet >_F g \longleftrightarrow f \bullet >_F g$ and Normal f(g, F). (That is, g is a normal form of f with respect to F).

$f \nabla_F^{succ} g \longleftrightarrow \exists h$ such that $f \bullet >_F h, g \bullet >_F h.$

$Spol(f, g) \triangleq Hcoef(g) \cdot \frac{LCM(Hterm(f), Hterm(g))}{Hterm(f)} \cdot f -$
$Hcoef(f) \cdot fz)|^P t \frac{LCM(Hterm(f), Hterm(g))}{Hterm(g)} \cdot g.$

If $F = \{f_1, f_2, \ldots, f_r\}$ is a finite sequence of polynomials, $\{j_1, j_2, \ldots, j_k\}$ are integers with $1 \leq j_i \leq L(F)$, then $H_F(\{j_1, j_2, \ldots, j_k\})$ denotes the LCM of the head terms of $f_{\{j_1}, f_{j_2}, \ldots, f_{j_k}\}.$

Consider the polynomial ring $R = \mathbf{K}[z_1, z_2, \ldots, z_n]$. A system of generators of an ideal $I \subset R$ is called a Gröbner basis if the leading monomials of its elements generate the same ideal in R as the leading monomials of all the polynomials in I. The notion of a Gröbner basis admits a number of equivalent reformulations and serve as the foundation of the whole of constructive commutative algebra. Let $R = Q[z_1, z_2, z_3]$, where Q denotes the field of rationals. Let $F = \{f_1, f_2\}$ denote a sequence of two polynomials,

$$f_1 = z_2^2 + z_2 z_1 + z_1^2, \quad f_2 = z_2^2 z_1 + 1.$$

Then, the polynomial $f = z_1 f_1 - f_2 \in Id(F)$, the polynomial ideal set up by the two polynomials in the sequence F of polynomials. However, $f = z_2 z_1^2 + z_1^3 - 1$ is in normal form modulo $\{f_1, f_2\}$ with respect to every term order since the head terms, z_2^2 and $z_2^2 z_1$ have been lifted to their least common multiple $z_2^2 z_1$ and subtracted so that the head monomials formed in f cancel out.

Gröbner basis algorithms are based on the significant fact that if the finitely many differences of the kind illustrated in the above example all reduce to zero then every polynomial in $Id(F)$ reduces to zero to make F a Gröbner basis, as defined next. The theorem below summarizes several equivalent definitions for Gröbner bases.

Let $F = \{f_1, f_2, \ldots, f_r\}$ be a sequence of polynomials. The following conditions are equivalent and each provides a definition for Gröbner bases.

1. $g \in Id(F) \longleftrightarrow g \bullet >_F 0$.
2. For $1 \leq i < j \leq L(F)$, $\mathrm{Spol}(f_i, f_j) \bullet >_F 0$.
3. $h \bullet >_F g, h \bullet >_F f \Rightarrow g = f$.
4. $\forall 1 \leq i < j \leq L(F)$, $\exists$ sequence $i = j_1, j_2, \ldots, j_k = j$, such that

$$H_F(j_1, j_2, \ldots, j_k) = H_F(i, j)$$

and

$$\mathrm{Spol}(f_{j_l}, f_{j(l+1)}) \bullet >0, l = 1, 2, \ldots, (k-1).$$

$f \bullet >_F^{(1)} g \Rightarrow (f+h) \nabla_F^{succ} (g+h)$.

Proof. $f \bullet >_{F,t,i}^{(1)} g \longleftrightarrow g = f - a \cdot s \cdot f_i$, where $a = \frac{Coef(t,f)}{Hcoef(f_i)}$, $s = \frac{t}{Hterm(f_i)}$.

- Case (a): $Coef(t, h) = 0$. Then, $f + h \bullet >_F^{(1)} g + h$.
- Case (b): $Coef(t, h) = -Coef(t, f)$. = Then, $Coef(t, f + h) = 0$ and $f + h \bullet >^{(1)} g + h$.
- Case (c): Neither of above. Let,

$$\hat{h} = f + h - \frac{Coef(t, f+h)}{Hoef(f_i)} \cdot s \cdot f_i, \tilde{h} = g + h - \frac{Coef(t, g+h)}{Hcoef(f_i)} \cdot s \cdot f_i.$$

Then, $\hat{h} = \tilde{h}$ and $f + h \, \nabla^1 g + h.$

$$f - g \bullet >_F 0 \;\Rightarrow\; f \, \nabla_F^{succ} g.$$

Proof. By induction on the number of $\bullet >^{(1)}$ steps needed to reduce $f - g$ to zero.

- Case (a): $f - g = 0$. Then M-reduction is possible in zero steps. Then $f - g$ and $f \, \nabla_F^{succ} g$.
- Case (b): $f - g + h_0 \nabla^{(1)} h_1 \nabla^{(1)} h_2 \nabla^{(1)} \ldots h_{t+1} = 0$.
  It is easy to show that $f - g \bullet >^{(1)} h_1 \;\Rightarrow\; \exists f^{(1)}, g^{(1)}$ such that

$$f \bullet > f^{(1)}, g \bullet > g^{(1)}, h_1 = f^{(1)} - g^{(1)}.$$

By induction hypothesis, $f^{(1)} \, \nabla_F^{succ} g^{(1)}$.
     Therefore, $f \, \nabla_F^{succ} g$.

Theorem 1.7. *The following conditions are equivalent:*

$$g \in \; < f_1, \ldots, f_r > \;\Rightarrow\; g \bullet >_F 0 \tag{1.23}$$

$$For \; 1 \leq i < j \leq L(F), \quad Spol(F_i, F_j) \bullet >_F 0 \tag{1.24}$$

$$h \bullet > h_1^{succ}, h \bullet > h_2^{succ} \;\Rightarrow\; h_1 = h_2 \tag{1.25}$$

$F_1, F_2, \ldots, F_r$ is called a Gröbner basis (for $< F_1, \ldots, F_r >$) if (1.23) is satisfied. If $F_1, \ldots, F_r$ is a Gröbner basis for the ideal $I = < F_1, \ldots, F_r >$, then to determine whether or not a given polynomial f is in I one M-reduces f to normal form which by (1.25) is unique; if this normal form is zero then f is clearly in I by the definition of M-reduction; if it is not zero then f is not in I by (1.23). To M-reduce a polynomial to normal form successively eliminate M-terms t from f by executing a step of the form $f \, \overset{\bullet >}{_{F,t,i}} g$ until no M-terms are left. To do this in the most efficient way, eliminate at each step the highest M-term t with respect to the ordering $< T$; it is clear that in a finite number of steps a normal form will be reached.

The above discussion, of course, assumes that $F_1, \ldots, F_r$ form a Gröbner basis for I. Given an arbitrary set of polynomials $F_1, \ldots, F_r$, the following algorithm gives a method for constructing a Gröbner basis $G_1, \ldots, G_m$ such that $< F_1, \ldots, F_r > = < G_1, \ldots, G_m >$.

Algorithm 1.1. *Let* $F_i = G_i, \; i = 1, \ldots, r$.

Initial conditions : $G = \{G_1, \ldots, G_r\}, B = \{(i,j) : 1 \leq i < j \leq L(G)\}$.

(We consider B as a set of unordered pairs). Suppose at a particular point in the algorithm, $G = G_1, \ldots, G_\ell, (\ell = L(G))$
    If $B = 0$ stop.
    If not, choose $(i,j) \in B$, form $Spol(G_i, G_j)$, reduce this to normal form with respect to G and denote the resulting polynomial by $G_{\ell+1}$. (Note that this polynomial is not necessarily unique because normal forms are not unique unless G is a

Gröbner basis). If $G_{\ell+1} \neq 0$ *redefine G to be the set* $G_1, \ldots, G_\ell, G_{\ell+1}$ *and B to be the set* $(B \setminus \{(i,j)\}) \cup \{(k, \ell+1) : 1 \leq k \leq \ell\}$ *and continue to the next step. If* $G_{\ell+1} = 0$ *just set* $B = \setminus \{(i,j)\}$ *and continue to the next step.*

This algorithm terminates in a finite number of steps (see proof below) and at the end one has a set of polynomials $\{G_1, \ldots, G_m\}$ which satisfy (1.24) and therefore form a Gröbner basis for

$$< G_1, \ldots, G_m > = < F_1, \ldots, F_r >$$

Note that by keeping track of the calculations one can find $A_{ij} \in \mathcal{K}[\underline{x}]$, $i = 1, \ldots, m, j = 1, \ldots, r$ such that

$$G_i = \sum_{j=1}^{r} A_{ij} F_j, \quad i = 1, \ldots, m. \tag{1.26}$$

Then, given a polynomial $f \in \mathcal{K}[\underline{x}]$, if we keep track of calculations in reducing f to normal form and we find that $f \bullet >_G 0$, we can compute $a_i \in \mathcal{K}[\underline{x}]$, $i = 1, \ldots, m$ such that $f = \sum a_i G_i$. (1.26) will then give us f in terms of the F_i's:

$$f = \sum_{i=1}^{m} \sum_{j=1}^{r} a_i A_{ij} F_j$$

The algorithm (1.1) can be refined to weed out certain unnecessary calculations and a strategy may be adopted for selecting pairs (i,j) from B. We state just one such modification (for proofs and further details see [24] and [25]).

Theorem 1.8. *At any step in the algorithm, given* $(i,j) \in B$, *if there exists u with* $1 \leq u \leq L(G)$ *and* $i \neq u \neq j$ *such that* $(i,u) \notin B$, $(u,j) \notin B$ *and* $Hterm(G_u)$ *divides* $LCM(Hterm(G_i), Hterm(G_j))$, *then*

$$Spol(G_i, G_j) \bullet >_G 0^{succ}.$$

Theorem 1.8 suggests (see [25, p. 12]) the strategy of choosing $(i,j) \in B$ such that $LCM(Hterm(G_i), Hterm(G_j))$ is a minimum with respect to the ordering $< T$.

Before we illustrate the construction of Gröbner bases with an example, we present the proof for the termination of the algorithm; this is done purely for the convenience of the reader because the original proof may be inaccessible to some.

Theorem 1.9. *Termination proof* (Buchberger [26, §3.2]).

In the algorithm, each time we add a new polynomial to G it is in normal form with respect to the existing polynomials in G. Therefore, if we consider the sequence of headterms of elements of G, it is clear that we will obtain (except for the first few headterms belonging to the initial polynomials) a sequence $t_1, t_2, t_3, \ldots$ of terms such that t_j is not a multiple of t_i for $j > i$. Buchberger calls such a sequence an M-sequence (M-Folge). The proof will be complete if we show that any M-sequence is finite. This is proved by induction on the number of variables n. If $n = 1$ the result is trivial. Assume it is true for $n < N$ and consider an M-sequence which begins with the term $x_1^{J_1} \cdots x_N^{J_N}$. If $x_1^{v_1} \cdots x_N^{v_N}$ is another element of the sequence then $v_i < I_i$ for at least one i $(1 \leq i \leq N)$.

For each k-tuple $(v_{i_1}, \ldots, v_{i_k}$ with $1 \leq k \leq N$, $1 \leq i_1 < i_2 < \cdots < i_k \leq N$ and $0 \leq v_{i_j} < I_i, j = 1, \ldots, k$, let $S(v_{i_1}, \ldots, v_{i_j})$ denote the subsequence of terms in the M-sequence such that the exponent of x_{i_j} is v_{i_j} , $j = 1, \ldots, k$, and the exponent of x_r is $\geq I_r$ for $r \notin \{i_1, \ldots, i_k\}$. Then it is clear that the M-sequence is partitioned into the $S(v_{i_1}, \ldots, v_{i_k})$'s which are finite in number.

If we consider a particular subsequence $S(v_{i_1}, \ldots, v_{i_k})$, the sequence obtained by eliminating $x_{i_1}, \ldots, x_{i_k}$ from each term is also an M-sequence because x_{i_j} has exponent v_{i_j} for every term in $S(v_{i_1}, \ldots, v_{i_k}), j = 1, \ldots, k$. Since $k \geq 1$ we conclude by the induction hypothesis that $S(v_{i_1}, \ldots, v_{i_k})$ is finite. The original M-sequence is therefore partitioned into a finite number of finite sums and consequently it is finite itself.

In the following example, $f_1 \xrightarrow[\bullet>]{-atG_i} g$ will represent $f\bullet >_G^{(1)} g$ and $g = f - atG_i$ where a is a constant and t is a term. Let

$$F_1(x_1, x_2) \ = G_1(x_1, x_2) \ = x_1^2 x_2^2 - 2x_1 x_2^2 - 2x_1 x_2 + 4x_1 + 4$$

$$F_2(x_1, x_2) \ = G_2(x_1, x_2) \ = x_1 x_2 - x_1 - 2x_2$$

$$spol(G_1, G_2) = G_1 - x_1 x_2 G_2 = x_1^2 x_2 - 2x_1 x_2 + 4x_1 + 4$$

$$\xrightarrow[\bullet >]{-x_1 G_2} x_1^2 + 4x_1 + 4^{succ} = G_3$$

Recall that G_3 is in normal form i.e. no term of G_3 is a multiple of the head terms of G_i, $i < 3$.

$$G_3 = G_1 - x_1(x_2 + 1)G_2$$

At this point $B = \{(1, 3), (2, 3)\}$. Select $(2, 3)$ because

$$LCM(x_1 x_2, x_1^2) \ <T \ LCM(x_1^2 x_2^2, x_1^2).$$

$$spol(G_2, G_3) = x_1 G_2 - x_2 G_3 = -6x_1 x_2 - x_1^2 - 4x_2$$

$$\xrightarrow[\bullet >]{+6G_2} -x_1^2 - 16x_2 - 6x_1$$

$$\xrightarrow[\bullet >]{+G_3} -16x_n x_1 + 4^{succ} = G_4$$

$$G_4 = (x_1 + 6)G_2 + (1 - x_2)G_3 = (1 - x_2)G_1 + (x_1 x_2^2 + 6)G_2$$

At this point $B = \{(1,3), (1,4), (2,4), (3,4)\}$; select $(2,4)$.

$$spol(G_2, G_4) = G_2 + \tfrac{x_1}{16}G_4 = \frac{1}{8}x_1^2 - 2x_2 - \frac{3}{4}x_1$$

$$+\tfrac{1}{8}G_3 \quad -2x_2 - \frac{1}{4}x_1 + \frac{1}{2}$$
$$\bullet >$$

$$-\tfrac{1}{8}G_4 \quad 0^{succ} = G_5$$
$$\bullet >$$

Now $B = \{(1,3), (1,4), (3,4)\}$; select $(3,4)$. $(3,2) \notin B$, $(2,4) \notin B$ and $LCM(HtermG_3, HtermG_4) = x_1^2 x_2$ which is divisible by $HtermG_2$, so by 1.8, $Spol(G_3, G_4)\bullet > 0^{succ}$.

Similarly $(1,2) \notin B$, $(2,3) \notin B$ and $LCM(HtermG_1, HtermG_3) = x_1^2 x_2^2$ which is divisible by $HtermG_2$, so $Spol(G_1, G_3)\bullet > 0^{succ}$. Finally $(1,2) \notin B$, $(2,4) \notin B$ and $HtermG_2$ divides $LCM(HtermG_1, HtermG_4)$ so $Spol(G_1, G_4)\bullet > 0^{succ}$.

At this point B is empty and the algorithm terminates. The basis generated is

$$G_1 = x_1^2 x_2^2 - 2x_1 x_2^2 - 2x_1 x_2 + 4x_1 + 4$$

$$G_2 = x_1 x_2 - x_1 - 2x_2$$

$$G_3 = x_1^2 + 4x_1 + 4$$

$$G_4 = -16x_n x_1 + 4$$

We could in fact delete G_1 and G_2 from the basis and remain with a Gröbner basis generating the same ideal because their headterms are multiples of other headterms in the basis (see [27, 28] for this and other results on minimality and uniqueness for Gröbner bases).

Continuing the example, we ask whether $(x_1 + 2)(x_1 + x_2 + 3)$ is in $< G_1, G_2 >$.

$$(x_1 + 2)(x_1 + x_2 + 3) = x_1^2 + x_1 x_2 + 2x_2 + 5x_1 + 6$$

$$\overset{-G_3}{\underset{\bullet >}{}} x_1 x_2 + 2x_2 + x_1 + 2$$

$$\overset{-G_2}{\underset{\bullet >}{}} 4x_2 + 2x_1 + 2$$

$$\overset{+\tfrac{1}{4}G_4}{\underset{\bullet >}{}} \frac{3}{2}x_1 + 2 \overset{succ}{\neq} 0$$

So, $(x_1 + 2)(x_1 + x_2 + 3) \notin\; < G_1, G_2 >$.

However, $(x_1 + 2)(2x_1 + 2x_2 + 3)$ is in $< G_1, G_2 >$ because

$$
\begin{aligned}
(x_1 + 2)(2x_1 + 2x_2 + 3) \;=\;& 2x_1^2 + 2x_1 x_2 + 4x_2 + 7x_1 + 6 \\[4pt]
\xrightarrow{\;\bullet>\;}^{-2G_3}\;& 2x_1 x_2 + 4x_2 - x_1 - 2 \\[4pt]
\xrightarrow{\;\bullet>\;}^{-2G_2}\;& 8x_2 + x_1 - 2 \\[4pt]
\xrightarrow{\;\bullet>\;}^{+\frac{1}{2}G_4}\;& 0^{succ}
\end{aligned}
$$

Furthermore

$$
(x_1 + 2)(2x_1 + 2x_2 + 3) = 2G_3 + 2G_2 + \frac{1}{2}G_4
$$

$$
\begin{aligned}
&= 2(G_1 - x_1(x_2 + 1)G_2) + 2G_2 - \frac{1}{2}((1 - x_2)G_1 \\
&\quad + (x_1 x_2^2 + 6)G_2) \\
&= \frac{1}{2}(3 + x_2)G_1 - \frac{1}{2}(x_1 x_2^2 + 4x_1 x_2 + 4x_1 + 2)G_2.
\end{aligned}
$$

We conclude this section by saying a bit about a vector version of Gröbner bases. Let

$$
e_i = \begin{bmatrix} 0 \\ \vdots \\ 0 \\ 1 \\ 0 \\ \vdots \\ 0 \end{bmatrix} \leftarrow i^{\text{th}}\; position\;\; e_i \in \mathcal{K}^m.
$$

We define a v-term (vector term) to be something of the form $e_i t$ where t is a scalar term in the previous sense and $1 \leq i \leq m$. We place an ordering on the v-terms as follows: $e_i t < T\; e_j s$ if $t < T\; s$ or when $t = s, i < j$.

Any vector polynomial $f \in \mathcal{K}^m[x]$ can be written uniquely as a linear combination of v-terms. We denote the highest term of $f \in \mathcal{K}^m[x]$ with respect to the above ordering by $Hterm(f)$ and its coefficient by $Hcoef(f)$. In general we denote the coefficient of a v-term τ $(=e_i t)$ in a vector polynomial f by $Coef(\tau, f)$. We say a v-term τ is a multiple of a v-term σ if $\tau = e_i t$, $\sigma = e_i s$ and t is a multiple of s - denote this by $Mult(\tau, \sigma)$. Define a reduction procedure with respect to a sequence

$$
F = \{F_1, F_2, \ldots, F_r\}, \;\; F_\ell \in \mathcal{K}^m[x]
$$

by $f \overset{\bullet>}{_{\tau,F,j}} {}^{(1)} g$ if τ occurs in f with non-zero coefficient,

$$F_j \qquad \neq \qquad 0$$

$$Mult(\tau, HtermF_j)$$

$$g \qquad = \qquad f - \frac{Coef(\tau,f)}{Hcoef(F_j)} \cdot \frac{t}{s} \cdot F_j$$

where $\tau = e_i t$, $HtermF_j = e_i s$.

As in the scalar case one can also introduce the notations $f\bullet >_F g$ and $f\bullet >_F g^{succ}$. Finally, let $f, g \in \mathcal{K}^m[\underline{x}]$, $Htermf = e_i t$, $Htermg = e_j s$ then

$$Spol(f,g) \triangleq \begin{cases} 0 & \text{if } i \neq j \\ Hcoef(g) \cdot \frac{LCM(s,t)}{t} \cdot f - Hcoef(f) \cdot \frac{LCM(s,t)}{s} \cdot g & \text{if } i = j \end{cases}$$

It is then interesting to note that the proofs of Bachmair and Buchberger for the scalar case go through mutatis mutandis for the vector case with the above definitions. In particular we have that 1.23, 1.24, and 1.25 are valid in the vector case where now $\{F_1, \ldots, F_r\}$ denotes the $\mathcal{K}[\underline{x}]$-submodule of $\mathcal{K}^m[\underline{x}]$ generated by $F_1, \ldots, F_r$. Also we can construct an algorithm similar to 1.1 which will terminate after a finite number of steps and yield a vector Gröbner basis. For further techniques involving computation with polynomial ideals, see Seidenberg [29].

1.6.2 Algorithms for Construction of Gröbner Bases

Problem Given a finite sequence of polynomials F, find another sequence of polynomials G such that,

Ideal(F) = Ideal (G), and G is a Gröbner basis.

Basic Algorithm
  Starting conditions: $G := F$, $B := \{(i,j) : 1 \leq i < j \leq L(F)\}$.
  While $\exists$ $(i,j) \in B$
  Do:

$$h := Spol(g_i, g_j) \; ; \; H = Normalf(h)$$

$$If \; h \neq 0, then \; G := (G,h) \, , B : B \bigcup \{(i, L(G) : 1 \leq< L_G - 1\}$$

$$B := B - \{I, J\}$$

Although it seems that in general one is adding more members to G and B as one progresses, one can in fact show that the algorithm terminates in a finite number of steps. At the termination of the algorithm, a Gröbner basis results.

Example 1.9. Let us continue the previous example with the lexicographic ordering,

$$1 < z_1 < z_2 < z_1^2 < z_1 z_2 < z_2^2 \,.$$

Here,

$$g_1 = f_1, \ \ g_2 = f_2 \,, \ g_3 = Spol(g_1, g_3) = z_2 z_1^2 + z_1^3 - 1.$$

$$Spol(g_1, g_3) = z_1^2 g_1 - z_2 g_3 = z_1^4 + z_2 \triangleq g_4.$$

$$Spol(g_2, g_3) = z_1 g_2 - z_2 g_3 = -z_1^3 z_2 + z_f z)|^P 1 + z_2.$$

$$Spol(g_2, g_3) + z_1 g_3 = z_1^4 + z_2.$$

$$Spol(g_1, g_4) = z_1^4 g_1 - z_2^2 g_4 = z_2 z_1^5 + z_1^6 - z_2^3.$$

Therefore,

$$Spol(g_1, g_4) \bullet >_{g_4} z_1^6 - z_2^3 \ - z_2^2 z_1 \bullet >_{g_4} - z_2^3 - z_2^2 z_1 \ - z_2 z_1^2 \bullet >_{g_1} 0.$$

It can be verified that g_1 , g_2 , g_3 , *and* g_4 form a Gröbner basis using arguments described next for a different ordering.

If the lexicographic ordering is changed to,

$$1 < z_2 < z_1 < z_2^2 < z_1 z_2 < z_1^2 \,,$$

then it can be shown that the Gröbner basis is composed of polynomials g_1 , g_2 , *and* $z_2^4 - z_1 - z_2$. Note that this is justifiable from the following calculations.

$$Spol(g_2, g_3) = z_2^2 g_2 - z_1 g_3 = z_1^2 + z_1 z_2 - z_2^2 \bullet >_{g_1} = 0.$$

$$Spol(g_1, g_3) = z_2^4 g_1 - z_1^2 g_3 \bullet >_{g_3} 0 \bullet >_{g_1} 0 \,.$$

Also, obviously

$$Spol(g_1, g_3) \bullet >_{g_3} 0 \,.$$

Example 1.10. It is given that the starting conditions for application of the algorithm for construction of the Gröbner basis are:

$$G : g_1 = f_1 = z_1^3 z_2 z_3 - z_1 z_3^2, g_2 = f_2 = z_1 z_2^2 z_3 - z_1 z_2 z_3, g_3 = f_3 = z_1^2 z_2^2 - z_3^2.$$

$$B : \{(1, 2), \ (1, 3), \ (2, 3)\}.$$

The lexicographic ordering is standard as in the beginning of this example. Therefore, the head terms of g_1, g_2, and g_3 are, respectively, $z_1^3 z_2 z_3$, $z_1 z_2^2 z_3$, and $z_1^2 z_2^2$.

$$Spol(g_2, g_3) = z_1 g_2 - z_3 g_3 = -z_1^2 z_2 z_3 + z_3^3 \triangleq -g_4.$$

Improved Form of Basic Algorithm For each $(i, j) \in B$, check to see if

$$\exists\, i = j_1 = j_2, \ldots, = j_k = j \text{ with } H_G(j_1, j_2, \ldots, j_k) = H_G(i, j)$$

and

$$\forall l, 1 \le l < k, (j_l, j_{l+1}) \notin B.$$

If such a sequence can be found, then (i, j) can immediately be eliminated from B and upon termination a Gröbner basis results. This algorithm suggests the strategy of choosing $(i, j) \in B$ with $H_G(i, j)$ minimal with respect to the ordering.

Example 1.11. For the lexicographic ordering induced by $z_1 < z_2$, consider the two starting polynomials,

$$g_1 = 2z_1^2 z_2 + 3z_1 + 1,$$
$$g_2 = 3z_2^2 z_1 + 2z_2 + 2.$$

$$
\begin{aligned}
3z_2 g_1 - 2z_1 g_2 &= 9z_1 z_2 + 3z_2 - 4z_1 z_n z_1 \\
&= \underline{5z_1 z_2 + 3z_2 - 4z_1} \\
&= g_3
\end{aligned}
$$

$$
\begin{aligned}
5 g_n z_1 g_3 &= 15z_1 + 5 - 6z_1 z_2 + 8z_1^2 \\
&\xrightarrow{+2g_3} \underline{4z_1 z_2 + 6z_2 + 8z_1^2 + 7z_1 + 5} \\
&= g_4
\end{aligned}
$$

$$
\begin{aligned}
4 g_3 - 5 g_4 &= 12z_2 - 16z_n z_2 - 40z_1^n z_1 - 25 \\
&= \underline{-18z_n z_1^2 - 51z_n} \\
&= g_5
\end{aligned}
$$

g_i's with Hterm dividing $z_1^2 z_2$:	g_1	g_3	g_4	g_5
Hcoef vector :	2	5	4	-18
Basis for vectors orthogonal to	0	4	-5	0
Hcoef vector	0	2	2	1
	5	-2	0	0
	-2	0	1	0

$$-2g_1 + z_1 g_4 \ = \ -6z_1 - 2 + 6z_1 z_2 + 8z_1^3 + 7z_1^2 + 5z_1$$

$$\xrightarrow{-2g_3 + g_4} \ \underline{8z_1^3 + 15z_1^2 + 14z_1 + 3}$$

$$= \ g_7$$

We now see $g_6(= -5g_7)$ is obsolete and we eliminate it from the basis.

g_i's with Hterm dividing $z_1^3 z_2$:	g_1	g_3	g_4	g_5	g_7
Hcoef vector :	2	5	4	-18	8
Basis for vectors orthogonal to	0	4	-5	0	0
Hcoef vector	0	2	2	1	0
	5	-2	0	0	0
	-2	0	1	0	0
	-4	0	0	0	1

$$-4z_1 g_1 + z_2 g_7 \ = \ -12z_1^n z_1 + 15z_1^2 z_2 + 14z_1 z_2 + 3z_2$$

$$\xrightarrow{-3z_1 g_3} \ -12z_1^2 - 4z_1 + 15z_1^2 z_2 + 14z_1 z_2 + 3z_n z_1^2 z_2$$

$$-9z_1 z_2 + 12$$

$$= \ 5z_1 z_2 + 3z_2 - 4z_1$$

$$\xrightarrow{-g_3} \ 0$$

g_i's with Hterm dividing $z_1 z_2^2$:	g_2	g_3	g_4	g_5
Hcoef vector :	3	5	4	-18
Basis for vectors orthogonal to	6	0	0	1
Hcoef vector	-3	1	1	0
	-4	0	3	0

$$-3g_2 + z_2 g_3 + z_2 g_4 = 9z_2^2 + 8z_1^2 z_2 + 3z_1 z_2 - z_2 - 6$$

$$\xrightarrow{-4g_1} 9z_2^2 + 8z_1^2 z_2 + 3z_1 z_2 - z_n - 8z_1^2 z_2$$

$$-12z_1 - 4$$

$$= 9z_2^2 + 3z_1 z_2 - z_n z_1 - 10$$

$$\xrightarrow{+g_n g_4} 9z_2^2 - 10z_n z_1^2 - 30z_n$$

$$= g_8$$

g_i's with Hterm dividing $z_1 z_2^2$:	g_2	g_3	g_4	g_5	g_8
Hcoef vector :	3	5	4	-18	9
Basis for vectors orthogonal to	-3	1	1	0	0
Hcoef vector	-4	0	3	0	0
	6	0	0	1	0
	-3	0	0	0	1

$$-4g_2 + 3z_2 g_4 = -8z_2 - 8 + 18z_2^2 + 24z_1^2 z_2 + 21z_1 z_2 + 15z_2$$

$$\xrightarrow{-2g_8} -8z_2 - 8 + 24z_1^2 z_2 + 21z_1 z_2 + 15z_2 + 20z_2$$

$$+32z_1^2 + 60z_1 + 40$$

$$= 24z_1^2 z_2 + 21z_1 z_2 + 27z_2 + 32z_1^2 + 60z_1 + 32$$

$$\xrightarrow{-12g_1} 21z_1 z_2 + 27z_2 + 32z_1^2 + 24z_1 + 20$$

$$\xrightarrow{-5g_3 + g_4} 18z_2 + 40z_1^2 + 51z_1 + 25$$

$$\xrightarrow{+g_5} 0$$

$$-3g_2 + z_1 g_8 = -6z_2 - 6 - 10z_1 z_n z_1^3 - 30z_1^n z_1$$

$$\xrightarrow{+2g_3} -16z_1^3 - 30z_1^n z_1 - 6$$

$$\xrightarrow{+2g_7} 0$$

$$6g_2 + z_1 z_2 g_5 = 12z_2 + nz_1^3 z_2 - 51z_1^2 z_n z_1 z_2$$

$$\xrightarrow{+5z_2 g_7} -51z_1^2 z_2 - 25z_1 z_2 + 12z_2 + 12 + 75z_1^2 z_2$$

$$+70z_1 z_2 + 15z_2$$

$$= \quad 24z_1^2 z_2 + 45z_1 z_2 + 27z_2 + 12$$

$$\xrightarrow{-12g_1} \quad 45z_1 z_2 + 27z_n z_1$$

$$\xrightarrow{-9g_3} \quad 0$$

g_i's with Hterm dividing $z_1^2 z_2^2$:	g_1	g_2	g_3	g_4	g_5	g_8
Hcoef vector :	2	3	5	4	-18	9
Basis for vectors orthogonal to	0	3	1	1	0	0
Hcoef vector	0	-4	0	3	0	0
	0	6	0	0	1	0
	0	-3	0	0	0	1
	0	0	4	-5	0	0
	0	0	2	2	1	0
	5	0	-2	0	0	0
	-2	0	0	1	0	0

g_i's with Hterm dividing $z_1^3 z_2$:	g_1	g_2	g_3	g_4	g_5	g_7	g_8
Hcoef vector :	2	3	5	4	-18	8	9
Basis for vectors orthogonal to	0	3	1	1	0	0	0
Hcoeff vector	0	-4	0	3	0	0	0
	0	6	0	0	1	0	0
	0	-3	0	0	0	0	1
	0	0	4	-5	0	0	0
	0	0	2	2	1	0	0
	5	0	-2	0	0	0	0
	-2	0	0	1	0	0	0
	-4	0	0	0	0	1	0

Groebner basis is:

$$g_1 = 2z_1^2 z_2 + 3z_1 + 1$$

$$g_2 = 3z_2^2 z_1 + 2z_2 + 2$$

$$g_3 = 5z_1 z_2 + 3z_2 - 4z_1$$

$$g_4 = 4z_1 z_2 + 6z_2 + 8z_1^2 + 7z_1 + 5$$

$$g_5 = 18z_2 + 40z_1^2 + 51z_1 + 25$$

$$g_7 = 8z_1^3 + 15z_1^2 + 14z_1 + 3$$

$$g_8 = 9z_2^n z_2 - 16z_1^2 - 30z_1 - 20$$

Note: 1.   We can eliminate g_1, g_2 from basis because any reduction in g_1, g_2 can be done via g_3, g_4.

2.   Can modify g_8:

$$g_8 \xrightarrow{+g_5} 9z_2^2 + 8z_2 + 24z_1^2 + 21z_1 + 5$$

Final reduced basis is:

$$h_1 = 5z_1z_2 + 3z_2 - 4z_1$$
$$h_2 = 4z_1z_2 + 6z_2 + 8z_1^2 + 7z_1 + 5$$
$$h_3 = 18z_2 + 40z_1^2 + 51z_1 + 25$$
$$h_4 = 8z_1^3 + 15z_1^2 + 14z_1 + 3$$
$$h_5 = 9z_2^2 + 8z_2 + 24z_1^2 + 21z_1 + 5$$

1.6.3   Properties and Some Uses

The solution of the problem posed in the previous subsection by application of the improved version of the basic algorithm eventually led to an algorithm referred to as the Gröbner-Buchberger (GB) algorithm [22]. This algorithm that computes a vector space over ground field K of the residue class ring

$$\frac{\mathbf{K}[z_1, z_2, \ldots, z_n]}{Ideal(F)},$$

where F is a finite subset of the multivariate polynomial ring $\mathbf{K}[z_1, z_2, \ldots, z_n]$, has the following properties.

1. Ideal (F) = Ideal GB(F).
2. GB(F) is a unique reduced Gröbner basis G of Ideal (F) with respect to the chosen term order.
3. GB makes possible effective computations in the residue class ring defined above.

The algorithm has many uses as explained though the examples below.

Example 1.12.   It is required to decide whether a specified Ideal (F) is principal. Proceed as follows

1. Apply the algorithm GB to compute $G = GB(F)$.
2. Then Ideal (F) is principal $= \iff$ G consists of exactly one element.

Example 1.13. It is required to decide whether a finite subset F of the multivariate polynomial ring $\mathbf{K}[z_1, z_2, \ldots, z_n]$ is solvable.

Proceed as follows

1. Apply the algorithm GB to compute $G = GB(F)$.
2. F is solvable $\iff 1 \notin G$.

Example 1.14. Given a finite subset F of the multivariate polynomial ring $\mathbf{K}[z_1, z_2, \ldots, z_n]$, it is required to compute the $i - th$ elimination ideal,

$$Ideal(F) \bigcap \mathbf{K}[z_1, z_2, \ldots, z_i].$$

Proceed as follows

1. Apply the algorithm GB to compute $G = GB(F)$, with respect to the chosen lexicographic ordering.
2. Then,

$$Ideal(F) \bigcap \mathbf{K}[z_1, z_2, \ldots, z_i] = Ideal(G \bigcap \mathbf{K}[z_1, z_2, \ldots, z_i]).$$

Example 1.15. It is required to enumerate all ideals in the multivariate polynomial ring $\mathbf{K}[z_1, z_2, \ldots, z_n]$, i. e.

find $F_1, F_2, \ldots, Ideal(F_i) \neq Ideal(F_j), \bigwedge_I \bigvee_j I = Ideal(F_j)$.
Proceed as follows

1. Enumerate all reduced GB

Example 1.16. It is required to determine whether or not a finite subset F of the multivariate polynomial ring $\mathbf{K}[z_1, z_2, \ldots, z_n]$ has infinitely zeros.

Proceed as follows

1. Compute $G = GB(F)$.
2. F has infinitely many zeros $\iff$ no polynomial of the form $z_i^j + \ldots \exists G$, for some i.

Example 1.17. Given a finite subset F of the multivariate polynomial ring $\mathbf{K}[z_1, z_2, \ldots, z_n]$, it is required to find B, the linearly independent basis for vector space, $\frac{\mathbf{K}[z_1, z_2, \ldots, z_n]}{Ideal(F)}$.

Proceed as follows:

1. Compute $G = GB(F)$.
2. $B = \{[z_1^{i_1} \ldots z_n^{i_n}] \mid z_1^{i_1} \ldots z_n^{i_n}$ is not a multiple of a head term of a polynomial in $G\}$.

Table 1.1 Summary of certain mathematical features that influence multidimensional signal processing

Item no.	Topics	Properties
1	Zero-set S of n-variate polynomial $p(z) = p(z_1, z_2, \cdots, z_n)$	$S = \{z = (z_1, z_2, \cdots, z_n) \mid p(z) = 0\}$ is unbounded when $n > 1$ and bounded when $n = 1$
2	n-variate holomorphic function	No holomorphic function has any isolated zero when $n > 1$. In particular, zeros of n-variate polynomials lie on continuous algebraic curves when $n > 1$ and are isolated when $n = 1$
3	n-variate irreducible rational functions	Have only nonessential singularities. Only in the case when $n > 1$, nonessential singularities of the second kind occur, which are always isolated and finite in number only when $n = 2$
4	Partial-fraction expansion of rational functions	Every n-variate rational function can be represented as a sum of fractions whose denominators contain at most n irreducible factors each. Construction of the representation is complicated due to the need for Hilbert's Nullstellensatz
5	Division algorithm and continued-fraction expansion	Holds in a principal ideal ring. The set of n-variate polynomials, when $n > 1$, does not constitute such a ring. The pseudo-division algorithm, however, enables the greatest common divisor to be constructed
6	n-dimensional residues	In the $n > 1$ case, the problem of computation of integrals of a closed holomorphic form over closed n-D surfaces cannot always be solved completely; sometimes it is only possible to reduce the dimension of the integral
7	Solution of n-D difference equations by the z-transform	In the $n > 1$ case, it is possible that a n-D difference equation with a specified initial set might have a unique solution, which might not be computable directly by the z-transform method
8	Convolutional solution for n-D difference equations	In the $n = 1$ case, a linear shift-invariant system characterized by a constant coefficient difference equation with zero initial conditions has a convolutional solution. In the $n > 1$ case, additional constraints on the system, besides linear shift-invariance, is necessary to guarantee this
9	Sum of squares representation of nonnegative definite forms	Under all nonnegative real forms of even order n in m variables, there are always some that cannot be written as a finite sum of squares of real forms. The only exceptions are the following three cases: (a) $n = 2$, m arbitrary (quadratic form), (b) $m = 2$, n arbitrary (binary form), and (c) $n = 4$, $m = 3$
10	Irreducible n-variate polynomials	When $n > 1$, almost all polynomials are irreducible
11	Classical n-D spectral factorization	When $n > 1$, finite order spectral factors do not exist, in general
12	Unisolvence of functions in $\mathbf{R}^n$	When $n = 1$, unisolvent systems are plentiful; when $n > 1$ the situation is vastly different

(continued)

Table 1.1 (continued)

Item no.	Topics	Properties
13	n-D state-space models	When $n > 1$, distinction has to be made between local and global state-space models. In general, the global state-space models are not finite-dimensional
14	Continuous positive-definite function in n-variables	For arbitrary n, every continuous positive-definite function is the Fourier transform of a finite positive measure. The converse assertion is also true
15	Fourier analysis over compact Abelian groups	As the torus T^n, the distinguished boundary of the unit polydisc U^n, is a compact Abelian group, an analysis that depends on group properties of a circle that generalize naturally

Chapter 2
Multivariate Polynomial Positivity (Nonnegativity) Tests

2.1 Introduction

In a variety of systems theory problems ranging from tests from Lyapunov stability, existence of limit cycles in nonlinear systems, existence of an operating point for a nonlinear circuit, the output feedback stabilization problem, multidimensional filter stability tests (see Chap. 4), tests for multivariate positive realness in electrical network realizability theory, etc., it is required to test a specified polynomial in several real variables for global or nonglobal positivity (nonnegativity). The topics of this chapter, dealing with the question of existence of such tests, followed by their actual construction depend a lot on the results of elementary decision algebra. Section 2.2 is, therefore, devoted to a concise exposition of the theory underlying the procedure for deciding the solvability in a real-closed field of a finite system of polynomial equations and inequalities with rational coefficients. For the development of this topic as a part of mathematical logic, the reader is referred to [30]. Here the discussion of the foregoing topic is centered around the quantifier elimination algorithm of Tarski [30] using a notion introduced by Cohen [31], the method of Seidenberg using the theory of resultants [32], and a cylindrical algebraic decomposition scheme for quantifier elimination developed by Collins in 1975 [2]. This last technique was also independently pursued by Bose and Modarressi, who used the theory of resultants to develop an algorithm for testing a multivariate polynomial for global positivity as reported in 1976 [2].

Section 2.3 is devoted to existing procedures for testing a specified multivariate polynomial with integer coefficients for global positivity (because then the test may be implementable with infinite precision if desired, subject to availability of sufficient computational resources, and also because the global positivity property of a polynomial with rational coefficients may be inferred with absolute certainty after performing a global positivity test on another polynomial with integer coefficients). The test based on the use of resultant-subresultants (inner determinants) to provide a computationally feasible quantifier elimination algorithm for real-closed fields

© Springer International Publishing AG 2017
N.K. Bose, *Applied Multidimensional Systems Theory*,
DOI 10.1007/978-3-319-46825-9_2

is amply illustrated by several nontrivial examples. The other tests considered are those which ultimately require a test for existence of a real solution to a number of multivariate polynomial equations in an equal number of indeterminates. Counterparts of all these tests, required to determine whether a specified multivariate polynomial having integer coefficients is globally nonnegative, are discussed in Sect. 2.4. Multivariate polynomial non-global positivity tests are considered in Sect. 2.5, and illustrative nontrivial examples are again given to familiarize the reader with the technicalities involved in actually carrying out such tests. In Sect. 2.6, the important problem of invariance of the positivity property under fluctuation of the original coefficient values is considered. Besides offering other benefits, this type of result will enable one to infer with absolute certainty the positivity property of a polynomial with real coefficients from a positivity test implementable with infinite precision on another polynomial with integer coefficients.

It is hoped that after studying this chapter the reader will be able to apply the procedures to solving physically motivated problems characterized by multivariate polynomials which must be tested for global or non-global positivity or non-negativity.

2.2   Elementary Decision Algebra

The decision problems for various mathematical theories and the related problems of definability have been of interest to mathematicians for more than half a century. The decision problem for a field is the problem of determining whether or not the set of true sentences of the field is recursive. If that set is recursive, the field is decidable; if not, the field is undecidable. The known decidable fields are: (a) any finite field; (b) any real-closed field; and (c) any algebraically closed field. Some of the known undecidable fields are the rational field, any finite extension of the rational field, and fields elementarily equivalent to any of these.

Among the known decidable fields, real-closed fields are decidable by Tarski's decision method, which gives a recipe for deciding in a finite number of steps the solvability of a finite system of multivariate polynomial equations and inequalities with rational coefficients. Tarski's result [30] is concerned with the question of how to decide elementary statements about real fields by means of a recursive procedure. It was shown that the theory of real-closed fields is decidable and complete. Before stating and proving Tarski's theorem, certain preliminaries required will be introduced next.

It is well known that in Sturm's theorem the number of distinct roots of a single-variable polynomial which fall within any given real interval may be determined through a process rational in the coefficients [33]. (Multiple roots can also be handled by such elementary devices as extraction of the greatest common factor from the given polynomial and its derivative, etc.) In 1922, Gummer presented a procedure using Sturmian sequences to determine the relative arrangement of roots for a system of univariate polynomials in any specified real interval, subject to the restrictions that the endpoints of the interval are not roots and that each root in the interval is of multiplicity one.

2.2.1 Real Root Ordering of Univariate Polynomial System

Consider two univariate polynomials $f(x_1)$ and $g(x_1)$. Let α denote the roots of $f(x_1)$, and β the roots of $g(x_1)$ in a given real interval (a, b). Assume that the endpoints for the interval are not roots of $f(x_1)$ and $g(x_1)$, and that all the real roots of $f(x_1)$ and $g(x_1)$ in (a, b) are simple roots. Writing down the roots within (a, b) in ascending numerical order, suppose one obtains $\alpha\alpha\alpha\beta\beta\alpha\beta\beta\beta\alpha\beta$. The β's effect a certain partitioning of the α's, which in this example is (3, 0, 1, 0, 0, 1, 0). The solution of the problem at hand consists in determining the number of α's in the successive groups. Gummer solved this problem by use of several Sturmian sequences generated from $f(x_1)$, $g(x_1)$, and their derivatives, and observance of sign variation and permanence of these sequences for values of $x_1 = a$ and $x_1 = b$. The details of the algorithm are quite cumbersome, and its main points are best illustrated through an example. It is assumed that $f(x_1)$ has no roots in (a, b) in common with any of the polynomials belonging to the Sturmian sequence generated from $g(x_1)$.

Example 2.1. Suppose the real roots of $f(x_1) = x_1^3 + 3x_1^2 - 3x_1 - 1$ and those of $g(x_1) = x_1^2 - 4x_1 - 7$ are to be ordered on the real line, i.e., $(a, b) = (-f, f)$, by rational operations. Let $g_0(x_1) = g(x_1)$, $f_0(x_1) = f(x_1)$, $g_1(x_1) = dg_0(x_1)/dx_1$, and $f_1(x_1) = df_0(x_1)/dx_1$. Here, $g_1(x_1) = 2x_1 - 4$ and $f_1(x_1) = 3x_1^2 + 6x_1 - 3$. Since the positive constants do not affect the sign, let $g_1(x_1) = x_1 - 2$ and $f_1(x_1) = x_1^2 + 2x_1 - 1$ be considered. The Sturmian sequence , $S[g_0(x_1), g_1(x_1)]$, for $g_0(x_1)$ is generated through a division process: $S[g_0(x_1), g_1(x_1)] = \{g_0(x_1), g_1(x_1), g_2(x_1)\}$, where $g_1(x_1) = x_1 - 2$, and $g_2(x_1) = +1$ (again the positive multiplicative constant dropped). Now, $G_1(x_1) = g_0(x_1).g_1(x_1)$, and $G_2(x_1) = g_1(x_1).g_2(x_1)$ are computed:

$$G_1(x_1) = (x_1^2 - 4x_1 - 7)(x_1 - 2) = x_1^3 - 6x_1^2 + x_1 + 14$$

$$G_2(x_1) = (x_1 - 2)(1) = x_1 - 2$$

To simplify, one can subtract from the G_i's, the product of any polynomial times $f_0(x_1)$. Denoting $G_1(x_1) - f_0(x_1)$ by $G_1(x_1)$, one obtains

$$G_1(x_1) = -9x_1^2 + 4x_1 + 15$$

$$G_2(x_1) = x_1 - 2$$

Also, $G_{12}(x_1) = G_1 G_2 = g_0 g_1 g_1 g_2 = g_0 g_1^2 g_2$ can be computed, and since g_1^2 does not contribute to its sign, it is taken as $G_{12}(x_1) = g_0(x_1) g_2(x_1)$:

$$G_{12}(x_1) = x_1^2 - 4x_1 - 7$$

Define, within the same sign for all x_1, $H_0(x_1) = f_1(x_1)$, $H_1(x_1) = G_1(x_1) f_1(x_1)$, $H_2(x_1) = G_2(x_1).f_1(x_1)$, and $H_{12}(x_1) = G_{12}(x_1).f_1(x_1)$. Here one has

$$H_0(x_1) = x_1^2 + 2x_1 - 1$$

$$H_1(x_1) = -17x_1^2 + 28x_1 - 1$$

$$H_2(x_1) = -3x_1^2 - 2x_1 + 3$$

$$H_{12}(x_1) = x_1^2 - 12x_1 + 1$$

It must be noted that the product of $f_0(x_1)$ and any polynomial can also be subtracted from the H_i's for the sake of degree reduction, without affecting the results. The Sturmian sequences $S[f_0(x_1), H_0(x_1)]$, $S[f_0(x_1), H_1(x_1)]$, $S[f_0(x_1), H_2(x_1)]$ and $S[f_0(x_1), H_{12}(x_1)]$ are now generated. For the example at hand, these are, within the same sign for all x_1,

$$S[f_0(x_1), H_0(x_1)] = \{(x_1^3 + 3x_1^2 - 3x_1 - 1), (x_1^2 + 2x_1 - 1), (x_1), (+1)\}$$

$$S[f_0(x_1), H_1(x_1)] = \{(x_1^3 + 3x_1^2 - 3x_1 - 1), (-17x_1^2 + 28x_1 - 1),$$
$$(-83x_1 + 23), (-1)\}$$

$$S[f_0(x_1), H_2(x_1)] = \{(x_1^3 + 3x_1^2 - 3x_1 - 1), (-3x_1^2 - 2x_1 + 3),$$
$$(8x_1 - 3), (-1)\}$$

$$S[f_0(x_1), H_{12}(x_1)] = \{(x_1^3 + 3x_1^2 - 3x_1 - 1), (x_1^2 - 12x_1 + 1),$$
$$(-11x_1 + 1), (+1)\}$$

If $V_{(i)}(a) = \text{var}\,\{S[f_0(a), H_{(i)}(a)]\}$, where var denotes the number of sign variations, and $M_{(I)} = V_{(I)}(a) - V_{(I)}(b)$, one has

$$V_0(-\infty) = 3, \; V_1(-\infty) = 2, \; V_2(-\infty) = 0, \; V_{12}(-\infty) = 1$$
$$V_0(\infty) = 0, \quad V_1(\infty) = 1, \quad V_2(\infty) = 3, \quad V_{12}(\infty) = 2$$

$$M_0 = V_0(-\infty) - V(\infty) = 3$$
$$M_1 = V_1(-\infty) - V_1(\infty) = 1$$
$$M_2 = V_2(-\infty) - V_2(\infty) = -3$$
$$M_{12} = V_{12}(-\infty) - V_{12}(\infty) = -1$$

$$M_0 = V_0(-\infty) - V(\infty) = 3$$
$$M_1 = V_1(-\infty) - V_1(\infty) = 1$$
$$M_2 = V_2(-\infty) - V_2(\infty) = -3$$
$$M_{12} = V_{12}(-\infty) - V_{12}(\infty) = -1$$

Now the distribution function $E(t)$ is formed, where

$$2^k E(t) = \sum_{(j)} M_{(j)}(1 + t)^{k-s}(1 - t)^s,$$

where k is the highest index of $g_i(x_1)$, and s is the number of elements in (j) (taken as 0 for M_0), e.g., $s = 2$ for M_{12}. For this example $k = 2$, and one finds

$$E(t) = \frac{1}{4}[3(1 + t)^2 + (1 + t)(1 - t) - 3(1 + t)(1 - t) - (1 - t)^2]$$

Arranging $E(t)$ in descending powers of t, one has

$$E(t) = t^2 + 2t.$$

The coefficients of powers of t in $E(t)$ give the number of α's between the consecutive β's, except for the outlying β's. Hence, so far one has $\alpha\beta\alpha\alpha$. To find the number of outlying β's one has to find γ and λ where,

$$\gamma = var\{S[g_0(a), g_1(a)]\}$$
$$\lambda = var\{S[g_0(b), g_1(b)]\}$$

Define γ' as the highest power of t in $E(t)$ and λ' as the lowest power of t in $E(t)$. For the example at hand, $\gamma = 2$, $\gamma' = 2$, $\lambda = 0$, and $\lambda' = 1$. Then the number of β's on either side of the string found from coefficients of t in $E(t)$ is given by $\gamma - \gamma'$ as the number of β's on the left-hand side of the string, and by $\lambda' - \lambda$ as the number of β's on the right-hand side of the string. Here one has

$$\gamma - \gamma' = 2 - 2 = 0$$
$$\lambda' - \lambda = 1 - 0 = 1.$$

Hence the final ordering is given by $\alpha\beta\alpha\alpha\beta$.

The procedure of Gummer was generalized to the case of multiple zeros by Meserve, using strategies exactly similar to those in Sturm's theorem for multiple zeros. Meserve also obtained other results related to the content of a finite system of univariate polynomial inequalities.

2.2.2 Tarski's Theorem in Elementary Decision Algebra

Several definitions are in order before the enunciation of the main theorem. The treatment closely parallels Cohen's development and proof [31] of Tarski's main result.

Definition 2.1. A set S is given with certain relations M_α defined on S. Each M_α is a subset of the direct product of S with itself N_α times for some integer N_α. An *elementary statement* (*sentence*) about the M_α is a statement formed by using the logical symbols denoting elementary operations on and elementary relations between real numbers: $\cap$ (conjunction, and); *bigcup* (disjunction, or); (negation, not); $\Rightarrow$ (implication); $\Leftrightarrow$ (equivalence); $=$ (equal); symbols for variables $x_1, x_2, \cdots$, which exclusively represent real numbers; universal and existential quantifiers, $\forall, \exists$, respectively; and the relation symbols M_α.

Algebraic equations and inequalities are among what are dubbed formulas of elementary algebra, and by combining equations and inequalities by logical expressions listed in Definition 2.1, arbitrary sentences of elementary algebra can be constructed. These sentences can, of course be true or false. Two sentences of elementary algebra follow; the first is true, but the second is false:

(a) For every $a_k, k = 0, 1, 2, \cdots, n$, with $a_n \neq 0$ and n an odd positive integer, there exists an x_1 such that $\sum_{k=0}^{n} a_k x_1^k = 0$.
(b) $0 > (1 + 1) + (1 + 1 + 1)$.

Definition 2.2. A decision method for a class S of sentence is a method by means of which it is always possible to decide in a finite number of steps whether any prescribed sentence S_1 belongs to the class S. A decision problem for a class S is the problem of finding or constructing a decision method for S.

Definition 2.3. A field **K** is called formally real if the only relations of the form $\sum_{k=0}^{m} a_k^2 = 0$ are those for which every $a_k = 0$, where $a_k \in K$ and m is a nonnegative integer.

The field of rational numbers, any field which is a purely transcendental extension of the field of rational numbers, and the field of rational functions in several indeterminates with coefficients in the field of rational numbers are all examples of formally real fields. It is simple to show that any field of nonzero characteristic, e.g., a finite field, cannot be formally real. Any formally real field can be ordered. The field axioms involve the following relations:

$$R_1(x_1, x_2, x_3) \equiv (x_1 + x_2 = x_3), \quad R_2(x_1, x_2, x_3) \equiv (x_1.x_2 = x_3), \qquad (2.1)$$

which follow the well known axioms concerning closure, associativity, commutativity, distributivity, inverses, identities, etc. A real-closed field, defined next, has more axioms.

Definition 2.4. A field **K** is called real-closed if **K** is formally real and no proper algebraic extension of **K** is formally real. (An algebraic extension is the field $K(\alpha)$ formed by adjoining a root of a polynomial with coefficients in **K**.)

The field of real numbers is a real-closed field. Any real-closed field can be ordered in one and only one way. The axioms of a real-closed field are essential to the understanding of the decision problem for real fields and are given next.

Fact 2.1. *The axioms of a real-closed field* **K** *are: (a) The axioms for a field. (b) The order axiom, which involves over and above the relations given in Eq. (2.1) a third relation*

$$R_3(x_1, x_2) \equiv (x_1 < x_2). \qquad (2.2)$$

The inequality in (2.2) represents an ordering, and $0 < x_1, 0 < x_2$ implies that $0 < x_1 x_2$, and $0 < x_1 + x_2$. Also, for any element x in K, one and only one of the relations $x - 0, x > 0, -x > 0$, holds. (c) The closure property, which states that for a m_1th degree polynomial, $(m_1 > 0), f(x_1) \in K[x_1]$, if given that $x_1 < x_2$ and $x_1, x_2 \in K$ it is true that $f(x_1) < 0$ and $f(x_2) > 0$, then $\exists x_3$ with $x_1 < x_3 < x_2 f(x_3) = 0$.

As K is real-closed, $K[\sqrt{-1}]$ is algebraically closed, so that $f(x_1)$ splits into linear and quadratic irreducible factors over K. Quadratic factors are of the form

$$x_1^2 + cx_1 + d = \left(x_1 + \frac{c}{2}\right)^2 + \left(d - \frac{c^2}{4}\right),$$

where because of irreducibility $d > c^2/4$. Therefore, sign changes come only from linear factors which go to zero between $x_1 = a$ and $x_2 = b$.

Definition 2.5. Let $x_1, x_2, \cdots, x_n$ be variables whose common domain is a Boolean algebra B. Then, a function $f(x_1, x_2, \cdots, x_n)$ built up from these variables and from elements of B by a finite number of application of the operations $\cap, \cup$, and negation is called a Boolean function of $x_1, x_2, \cdots, x_n$.

Definition 2.6. A polynomial relation $P(x_1, x_2, \cdots, x_n)$ is a Boolean function of a finite number of relations of the form

$$a(x_1, x_2, \cdots, x_n) > 0, \text{ where } a(x_1, x_2, \cdots, x_n) \in Z[x_1, x_2, \cdots, x_n],$$

where Z is the ring of integers. (In the original presentation of Tarski, relations of the form $a(x_1, x_2, \cdots, x_n) = 0$, were also included.)

Tarski's decision method is based on a procedure for eliminating quantifiers and therefore the observation summarized next is important, though not difficult to confirm.

Fact 2.2. *All elementary statements occurring in the theory of decision problem for real fields can be reduced to the form, $(Q_1 x_1)(Q_2 x_2) \cdots (Q_i x_i) P(x_1, x_2, \cdots, x_n)$ (a standard prenex formula), where the Q_i's are the universal or existential quantifiers $(i \le n)$, and $P(x_1, x_2, \cdots, x_n)$ is a polynomial relation and is quantifier-free.*

Fact 2.3. *It is possible to obtain, via use of rational operations only, the ordering of the zeros in any real interval of a system of univariate polynomial equations, irrespective of the multiplicity of the roots present.*

Cohen's proof of Tarski's theorem on decision procedure for real fields dwells heavily on the concept of effective real-valued functions, which is considered next.

Definition 2.7. A real-valued function $f(x_1, x_2, \cdots, x_n)$ is effective if there is a primitive recursive procedure which to every polynomial relation $P_1(y, x_{n+1}, x_{n+2}, \cdots, x_m)$ assigns a polynomial relation $P_2(x_1, x_2, \cdots, x_n, x_{n+1}, x_{n+2}, \cdots, x_m)$ such that

$$P_1(f(x_1, x_2, \cdots, x_n), x_{n+1}, \cdots, x_m) \leftrightarrow P_2(x_1, x_2, \cdots, x_n, x_{n+1}, \cdots, x_m).$$

The notion of primitive recursive procedure required in the above discussion is very briefly discussed next. For additional details, the reader is referred to standard texts on the theory of computation [34]. Functions computable by programs for a universal calculator are called partial functions.

Definition 2.8. Let n be a nonnegative integer, let g be a partial function of n variables (a function of zero variables is a constant), and let h be a partial function of $n + 2$ variables. A partial function f of $n + 1$ variables is said to be defined by primitive recursion from g and h provided that

$$f(x_1, x_2, \cdots, x_n, 0) = g(x_1, \cdots, x_n)$$

$$f(x_1, \cdots, x_n, x_{n+1} + 1) = h(x_1, \cdots, x_n, x_{n+1}, f(x_1, \cdots, x_{n+1})). \qquad (2.3)$$

These recursion equations determine $f(x_1, \cdots, x_n, x_{n+1})$ if the following conditions are satisfied: (i) If $f(a_1, \cdots, a_n, a_{n+1})$ is defined, then so is $f(a_1, \cdots, a_n, b)$ for all $b \leq a_{n+1}$ and all instances of the recursion equation corresponding to values $b \leq a_{n+1}$ are satisfied by f; (ii) f is undefined in all other cases.

Some well known primitive recursive (and therefore computable) functions are sum, product, power, modified difference (defined for x_1, x_2 as $|x_1 - x_2|$), and signum function (denoted as $\mathrm{sgn} x_1$, which has values -1, 0, 1 according as $x_1 < 0, x_1 = 0, x_1 > 0$, respectively). In Definition 2.7, effective functions, like computable functions, are closed under composition.

Fact 2.4. *$f(x_1, \cdots x_n)$ is effective if there is a primitive recursive function which assigns to every m a polynomial relation $P(c_0, \cdots, c_m, x_1, \cdots, x_n, y)$ such that*

$$P(c_0, \cdots, c_m, x_1, \cdots, x_n, y) \leftrightarrow y = \mathrm{sgn}(c_m f^m + \cdots + c_0).$$

Definition 2.9. Let $a(x_1)$ be a polynomial in a single indeterminate. By a graph for $a(x_1)$ is meant a k-tuple $x_{11} < x_{12} < \cdots < x_{1k}$ such that in each open interval of the form $(-\infty, x_{11}), (x_{11}, x_{12}), \cdots, (x_{1k}, \infty), a(x_1)$ is monotonic. By the data of the graph is meant the k-tuple $(x_{11}, \cdots, x_{1k})$ along with $\mathrm{sgn}(a(x_{1i})), i = 1, 2, \cdots, k$, $\mathrm{sgn}(a(x_{11} - 1))$, and $\mathrm{sgn}(a(x_{1k} + 1))$.

The following lemmas are required for proof of the main theorem to follow.

Lemma 2.1. *A graph for the univariate polynomial*

$$a(x_1) = \sum_{k=0}^{m_1} a_k x_1^k$$

is defined by $2m_1$ effective functions of the coefficients a_k's, namely x_{1k}, $\mathrm{sgn} a(x_{1k})$ for $k = 1, 2, \cdots, m_1 - 1$, $\mathrm{sgn} a(x_{11} - 1)$, and $\mathrm{sgn}(x_{1(m_1-1)} + 1)$, with $x_{11} < x_{12} < \cdots < x_{1(m_1-1)}$ forming a graph for $a(x_1)$.

Lemma 2.2. *For the univariate polynomial*

$$a(x_1) = \sum_{k=0}^{m_1} a_k x_1^k$$

there are $m_1 + 1$ effective functions of the a_k's, namely r and $b_{11} < b_{12} < \cdots < b_{1m_1}$, such that $b_{11}, b_{12}, \cdots, b_{1r}$ are all the (real) roots of $a(x_1)$.

Proof. The proofs for the above Lemmas 2.1 and 2.2 rest on the principle of mathematical induction. When $m_1 = 1$, the lemmas are trivially true. Now, assume that both lemmas are true for all values greater than 1 but less than a given m_1. To prove Lemma 2.1, consider the derived polynomial $a'(x_1) = da(x_1)/dx_1$, whose real zeros lie among $b_{11}, b_{12}, \cdots, b_{1(m_1-1)}$, which by Lemma 2.2 (as the degree of $da(x_1)/dx_1$ is less than m_1) are effective functions of the coefficients of $a(x_1)$, and effective functions are known to be closed under composition). By a similar reasoning, as sgn x_1 is an effective function, and b_{1k}, $1 \leq k \leq m_1 - 1$ are effective functions of the coefficients of $a(x_1)$, the functions sgn $a(b_{1k})$, $1 \leq k \leq m_1 - 1$, sgn $a(b_{11} - 1)$, and sgn $a(b_{1(m_1-1)} + 1)$ must be effective. Therefore the $2m_1$ effective functions of the a_k's, namely b_{1k}, sgn $a(b_{1k})(1 \leq k \leq m_1 - 1)$, sgn $a(b_{11} - 1)$, and sgn $a(b_{1(m_1-1)} + 1)$, form the data of a graph for $a(x_1)$.

To prove Lemma 2.2, let $x_{11} < x_{12} < \cdots < x_{1(m_1-1)}$ be effective functions of the a_k's defining a graph for $a(x_1)$, as is known to be possible from proved Lemma 2.1. By examining sgn $a(x_{1k})$, $1 \leq k \leq m_1 - 1$, sgn $a(x_{11} - 1)$, and sgn $a(x_{1(m_1-1)} + 1)$ it is possible to determine the number of real roots of $a(x_1)$ (this is also known to be possible by rational operations on the coefficients of $a(x_1)$ via use of Sturm's theorem [35]). In each of the open intervals $(-\infty, x_{11}), (x_{11}, x_{12}), \cdots, (x_{1(m_1-1)}, \infty)$, there can be at most one real root of $a(x_1)$, and it is required to prove that these roots are effective functions of the coefficients of $a(x_1)$. Consider the case of one possible root at $x_1 = b_{1i}$ in the interval $(x_{1i}, x_{1(i+1)})$. Before use of Fact 2.4, it is sufficient to show that if a polynomial is given to be of the form

$$c(x_1) = \sum_{k=0}^{m} c_k x_1^k,$$

then sgn $c(b_1I)$ is an effective function of the a_k's and c_k's. Based on arguments involving division of polynomials, clearly only the case $m < m_1$ need be considered. Then, invoking Fact 2.3, it is possible via rational operations to order the real roots of $c(x_1)$ and $a(x_1)$. This will lead to the determination of sgn $c(b_{1l})$, which therefore must be an effective function of the c_k's and a_k's. Consequently, by Fact 2.4 the root $x_1 = b_{1l}$ must be an effective function of the a_k's. Similar arguments can be advanced for any other root of $a(x_1)$, and the proof of Lemma 2.2 is now complete. Tarski's theorem is stated next.

Theorem 2.1. *If $P(x_1, x_2, \cdots, x_n)$ is a polynomial relation, $n > 1$, then it is possible to find via a primitive recursive procedure a polynomial relation $T(x_2, \cdots, x_n)$ such that*

$$(\exists x_1) P(x_1, x_2, \cdots, x_n) \leftrightarrow T(x_2, x_3, \cdots, x_n) \tag{2.4}$$

If $n = 1$, there is a primitive recursive procedure which decides, $(\exists x_1) P(x_1)$.

Proof. From Definition 2.6 it is evident that $P(x_1, x_2, \cdots, x_n)$ is a Boolean function of a finite number of relations of the form $a_I(x_1, x_2, \cdots, x_n) > 0$. The polynomial $a_I(x_1, x_2, \cdots, x_n)$ can be written in recursive canonical form in the main variable x_1. Then from Lemma 2.2 for any fixed $(n-1)$-tuple $(x_{20}, x_{30}, \cdots, x_{n0})$ the roots of $a_I(x_1, x_{20}, \cdots, x_{n0})$ are effective functions of $x_{20}, \cdots, x_{n0}$, and by Lemma 2.1 an effective graph for $a_I(x_1, x_{20}, \cdots, x_{n0})$ can be found. Therefore, by examining the various possibilities associated with the $(n-1)$-tuple $(x_2, x_3, \cdots, x_n)$, one can determine what the various possibilities are for the sequence $\{\operatorname{sgn} a_I(x_1, x_2, \cdots, x_n)\}$ for an arbitrary n-tuple $(x_1, x_2, \cdots, x_n)$. This in turn implies that a polynomial relation $T(x_2, x_3, \cdots, x_n)$, satisfying (2.4) can be found.

Theorem 2.1 summarizes a quantifier-elimination algorithm. From it follows the fact that the algorithm may accept as input any standard prenex formula of the form

$$(Q_1 x_1)(Q_2 x_2) \cdots (Q_i x_i) P(x_1, x_2, \cdots, x_n)$$

and supply as output an equivalent standard quantifier-free formula $T(x_{i+1}, x_{i+2}, \cdots, x_n)$. Though the proof of Tarski's theorem as outlined above is conceptually neat and elegant, its actual implementation is likely to be computationally involved, as is the case with Tarski's original proof. A simple yet nontrivial example illustrating this fact can be found in [36–39]. In fact, it can be shown that if r is the number of multivariate polynomials occurring in a prenex input formula and if m is the maximum degree of any such polynomial in any variable, then the maximum computing time needed to implement the quantifier-elimination algorithm of Tarski is exponential in both m and r, for a fixed n.

2.2.3 Seidenberg's Procedure

This procedure also works by successive reduction of the number of variables. However, the multivariate polynomial equality/inequality sets have a different form. Consider the following set of polynomials, where **x** for brevity denotes the n-tuple of variables $(x_1, x_2, \cdots, x_n)$ and the polynomials belong to $Q[x_1, x_2, \cdots, x_n]$ (Q, here, is the field of rational numbers):

$$f_\alpha(\mathbf{x}), \quad \alpha = 1, \cdots, r_\alpha; \quad g_\beta(\mathbf{x}), \quad \beta = 1, \cdots, r_\beta$$

$$h_\gamma(\mathbf{x}), \quad \gamma = 1, \cdots, r_\gamma; \quad k_\delta(\mathbf{x}), \quad \delta = 1, \cdots, r_\delta. \tag{2.5}$$

The decidability question is: "Is there a real n-tuple $\mathbf{x}$ in a real-closed field $\mathbf{K}$ containing Q which satisfies

$$f_\alpha(\mathbf{x}) = 0, \quad g_\beta > 0, \quad h_\gamma(\mathbf{x}) \geq 0, \quad k_\delta(\mathbf{x}) \neq 0, \tag{2.6}$$

where the subscripts $\alpha, \beta, \gamma, \delta$ range over all the associated integers indicated in (2.5)?" Obviously, one has

$$k_\delta(\mathbf{x}) \neq 0 \leftrightarrow k_\delta^2(\mathbf{x}) > 0. \tag{2.7}$$

Also, it is clear that

$$h_\gamma(\mathbf{x}) \geq 0 \leftrightarrow h_\gamma(\mathbf{x}) = 0 \, \text{or} \, h_\gamma(\mathbf{x}) > 0 \tag{2.8}$$

$$f_\alpha(\mathbf{x}) = 0, \alpha = 1, 2, \cdots, r_\alpha \leftrightarrow \sum_{\alpha=1}^{r_\alpha} f_\alpha^2(\mathbf{x}) = 0 \tag{2.9}$$

$$g_\beta(\mathbf{x}) > 0 \leftrightarrow x_{n+1}^2 g_\beta(\mathbf{x}) = 1. \tag{2.10}$$

In (2.10), x_{n+1} is another indeterminate. Using (2.7), (2.8), and (2.9), it is seen that $\mathbf{x}$ satisfies (2.6) if and only if it satisfies at least one of a number of sets of equation-inequalities of the following type:

$$f(\mathbf{x}) = 0, \quad g_\beta(\mathbf{x}) > 0, \quad \beta = 1, 2, \cdots, r_{(1)\beta}. \tag{2.11}$$

(Each inequality $h_{gamma}(\mathbf{x}) \geq 0$ doubles the number of such sets.) Using (2.9), (2.10) and (2.11) it is evident that $\mathbf{x}$ satisfies (2.6) if $(\mathbf{x}, x_{n+1}, \cdots, x_{n+n_1})$ satisfies any one of a number of polynomial equations of the type

$$f_{\epsilon\kappa}(\mathbf{x}, x_{n+1}, \cdots, x_{n+n_1}) = 0, \quad n_1 \equiv r_{(1)\beta}. \tag{2.12}$$

In (2.12), $x_{n+1}, \cdots, x_{n+n_1}$ are n_1 additional indeterminates necessitated by equivalences of the type (2.10). In (2.12), if $1 \leq k \leq n_\epsilon$ then $\mathbf{x}$ satisfies (2.6) if and only if $(\mathbf{x}, x_{n+1}, \cdots, x_{n+n_1})$ satisfies

$$\prod_{k=1}^{n_\epsilon} f_{\epsilon\kappa}(\mathbf{x}, x_{n+1}, \cdots, x_{n+n_1}) = 0. \tag{2.13}$$

As

$$g_\beta(\mathbf{x}) > 0 \leftrightarrow -g_\beta(\mathbf{x}) < 0 \tag{2.14}$$

and

$$h_\gamma(\mathbf{x}) \geq 0 \leftrightarrow -h_\gamma(\mathbf{x}) \leq 0, \tag{2.15}$$

the following statement is true:

Fact 2.5. *The answer to a decidability question on the existence of an n-tuple in a real-closed field satisfying any n-variate polynomial equality/inequality set involving signs,* $=, <, >, \geq, \leq,$ *and* $\neq$ *can be obtained by answering the decidability question on the existence of an* $(n+n_1)$*-tuple,* $n_1 \geq 0$*, in a real-closed field satisfying a single* $(n + n_1)$*-variate polynomial equation.*

Seidenberg's starting point is the replacement of an initially prescribed equality/inequality set with a single equality. It must, however, be borne in mind that from the computational standpoint this strategy might not be the most efficient. Seidenberg also gives a procedure for concluding for a prescribed pair of n-variate polynomials $f_n(\mathbf{x}), g_n(\mathbf{x})$ belonging to $Q[x_1, x_2, \cdots, x_n]$ (where Q may denote other formally real fields besides the field of rational numbers and g_n may be absent) that

$$f_n(\mathbf{x}) = 0, \quad g_n(\mathbf{x}) \neq 0 \qquad (2.16)$$

holds for some real n-tuple $\mathbf{x}$ if and only if one of a finite set of equations of the type

$$f_{n-1}(x_1, \cdots, x_{n-1}) = 0, \quad g_{n-1}(x_1, \cdots, x_{n-1}) \neq 0 \qquad (2.17)$$

holds for some real $(n-1)$-tuple $(x_1, \cdots, x_{n-1})$. The sets of f_{n-1}, g_{n-1} in (2.17) are computable from f_n and g_n in (2.16) by rational operations (involving operations of addition, subtraction, multiplication, and division only) and f_{n-1}, g_{n-1} belong to $Q[x_1, x_2, \cdots, x_{n-1}]$. Further, to within an inessential bijective transformation, if $(x_{10}, \cdots, x_{(n-1)0})$ is a solution to (2.17), then there is a solution to (2.16) of the form $(x_{10}, \cdots, x_{n0})$. This means that, knowing a particular solution to any of the sets in (2.17), one can construct a solution of (2.16) by factoring a univariate polynomial $f_n(x_{10}, \cdots, x_{(n-1)0}, x_n)$ in the variable x_n and checking which real zero $x_n = x_{n0}$ yields $g_n(x_{10}, \cdots, x_{n0}) \neq 0$.

Seidenberg proceeded by first giving a decision procedure for determining whether a bivariate polynomial $f(x_1, x_2) \in Q[x_1, x_2]$ has a zero in a real-closed field $\mathbf{K}$ containing Q (the case when $\mathbf{K}$ is the field of real numbers is of immediate interest to us). The procedure is based on the following facts.

Fact 2.6. *Let* $f(x_1, x_2) \in Q[x_1, x_2] \subset K[x_1, x_2]$*. Then, if* $f(x_1, x_2) = 0$ *has a solution in* $\mathbf{K}$*, it has a solution in* $\mathbf{K}$ *with* $x_1^2 + x_2^2$ *minimum; i.e., there is a pair* (x_{10}, x_{20}) *in* $\mathbf{K}$ *satisfying* $f(x_1, x_2) = 0$ *which is nearest to the point* $(0, 0)$*.*

Fact 2.6 can almost immediately be established when $\mathbf{K}$ is the field of real numbers; for proof of its validity when $\mathbf{K}$ is an arbitrary real-closed field, see [33, pp. 300–303], or [32, pp. 370–371]. Prior to stating Fact 2.7, it is necessary to define the notion of simple and singular points on plane algebraic curves.

Definition 2.10. A point (x_{10}, x_{20}) on $f(x_1, x_2) = 0$ is simple if and only if

$$\left(\left(frac{\partial f(x_1, x_2)}{\partial x_1} \right)_{(x_{10}, x_{20})}, \left(\frac{\partial f(x_1, x_2)}{\partial x_2} \right)_{(x_{10}, x_{20})} \right) \neq (0, 0).$$

Otherwise, the point (x_{10}, x_{20}) will be called singular.

Fact 2.7. *Let $f(x_1, x_2) \in Q[x_1, x_2] \subset K[x_1, x_2]$. Let $(x_{10}, x_{20}) \in (K, K)$ be a point of intersection of $f(x_1, x_2) = 0$ and a circle. Assume that (x_{10}, x_{20}) is simple and that the tangent to $f(x_1, x_2)$ at (x_{10}, x_{20}) has points interior to the circle. Then $f(x_1, x_2) = 0$ has points interior to the circle.*

The proof of Fact 2.7 can be found in [33, pp. 302–303], [32, p. 371]. The tangent line to the curve $f(x_1, x_2) = 0$ at a simple point (x_{10}, x_{20}) is given by

$$\left(\frac{\partial f}{\partial x_1}\right)_{(x_{10}, x_{20})} (x_1 - x_{10}) + \left(\frac{\partial f}{\partial x_2}\right)_{(x_{10}, x_{20})} (x_2 - x_{20}) = 0. \tag{2.18}$$

Fact 2.8, to be stated next, is trivially true when (x_{10}, x_{20}) is a singular point or when (x_{10}, x_{20}) coincides with the center, (x_{11}, x_{21}).

Fact 2.8. *Let $f(x_1, x_2) \in Q[x_1, x_2]$. If $f(x_1, x_2) = 0$ has a solution in K, then the equations*

$$f(x_1, x_2) = 0$$

and

$$g(x_1, x_2) = (x_2 - x_{21})\frac{\partial f}{\partial x_1} - (x_1 - x_{11})\frac{\partial f}{\partial x_2} = 0$$

have a common solution in K for any $x_{11}, x_{21} \in K$.

Fact 2.8 provides the key to determining whether or not $f(x_1, x_2) = 0$ has a solution in K. When K is the real-closed field of real numbers, the following approach for determining whether or not $f(x_1, x_2), g(x_1, x_2)$ has a common solution in K is almost self-evident. Without any loss of generality, $f(x_1, x_2)$ and $g(x_1, x_2)$ will be assumed to be relatively prime (a common factor, if present, could be similarly treated, after extraction, as the given polynomial $f(x_1, x_2)$). Let $r_1(x_1), r_2(x_1)$ be, respectively, the resultants after writing the two bivariate polynomials $f(x_1, x_2)$ and $g(x_1, x_2)$ in recursive canonical form, first with x_2 as main variable and then with x_1 as main variable. Since $r_1(x_1) \equiv 0(r_2(x_2) \equiv 0)$ if and only if $f(x_1, x_2)$ and $g(x_1, x_2)$ have a common factor of positive degree in $x_2(x_1)$, here by assumption, $r_1(x_1) \not\equiv 0, r_2(x_2) \not\equiv 0$. It is simple to establish that common zeros of $f(x_1, x_2), g(x_1, x_2)$ in K can only come from the zeros of $r_1(x_1)$ and $r_2(x_2)$ in K. By Sturm's theorem it is possible to determine within arbitrary accuracy the finite set of pairs comprised of all possible combinations of zeros in K—one of $r_1(x_1)$ and the other of $r_2(x_2)$. Each such pair can then be substituted to check whether it is a point on $f(x_1, x_2) = 0, g(x_1, x_2) = 0$. Other variants in the use of Sturm's theorem to decide whether $f(x_1, x_2)$ has a zero in K via use of Fact 2.8 can be found in [32, pp. 366–368].

Before extending the decidability problem under discussion to polynomials in more than two indeterminates, Seidenberg gave a procedure to decide for the existence of solutions in K for $f_1(x_1, x_2) = 0, g(x_1) \neq 0$ in terms of existence of solutions in $\mathbf{K}$ for $f(x_1, x_2) = 0$, where $f_1(x_1, x_2), f(x_1, x_2) \in Q[x_1, x_2] \subset K[x_1, x_2]$ and $g(x_1) \in Q[x_1] \subset K[x_1]$. This was done as follows: Without any loss of generality, $f_1(x_1, x_2)$ and $g(x_1)$ can be taken to be relatively prime (if not, the g.c.f. can be extracted from $g(x_1)$ and the polynomial coefficients in x_1 of $f_1(x_1, x_2)$, after writing it in recursive canonical form with x_2 as main variable). Also, assume that $f_1(x_1, x_2) = 0$ and $g(x_1) = 0$ do not meet on line $x_2 = 0$. Otherwise choose as the new x_1-axis the line $x_2 = x_{21}$, where x_{21} is greater than any real root of the resultant $r(x_2)$ of $g(x_1)$ and $f_1(x_1, x_2)$ written in recursive canonical form with x_1 as main variable). Define

$$f(x_1, x_2) \equiv f_1(x_1, x_2 g(x_1)) \tag{2.19}$$

Then, if a pair (x_{10}, x_{20}) satisfies $f(x_1, x_2) = 0$, then $f_1(x_{10}, x_{20} g(x_{10})) = 0$, $g(x_{10}) \neq 0$, because $f_1(x_1, 0)$ and $g(x_1)$ are relatively prime. Again if a pair (x_{10}, x_{20}) satisfies $f_1(x_1, x_2) = 0$, $g(x_1) \neq 0$, then $f(x_{10}, x_{20}[g(x_{10})]^{-1}) = 0$. These results are summarized in the following lemma.

Lemma 2.3. *Let $g(x_1) \in Q[x_1] \subset \mathbf{K}[x_1]$, $f_1(x_1, x_2) \in Q[x_1, x_2] \subset K[x_1, x_2]$. Without loss of generality, assume that $g(x_1)$ and $f_1(x_1, x_2)$ are relatively prime (of course, each can be assumed to be of nonzero degree in x_1, to avoid the trivial case), and furthermore assume that $f_1(x_1, 0)$ and $g(x_1)$ are relatively prime. Then $f_1(x_1, x_2) = 0$, $g(x_1) \neq 0$ has a solution in $\mathbf{K}$ if and only if $f(x_1, x_2) = 0$ (with $f(x_1, x_2)$ defined in (2.19)) has a solution in $\mathbf{K}$.*

The above scheme can be extended to apply to multivariate polynomial equalities/inequalities by treating all but two of the indeterminates as parameters, where each parameter belongs to a real-closed field. This extended result is summarized in the following theorem.

Theorem 2.2. *Let $f(t_1, t_2, \cdots, t_r; x_1, x_2) \in Q[t_1, t_2, \cdots, t_r; x_1, x_2]$ and $g(t_1, t_2, \cdots, t_r; x_1) \in Q[t_1, t_2, \cdots, t_r; x_1]$, where Q is the field of rational numbers. Then, it is possible to determine in a finite number of steps involving rational operations a finite set of pairs of polynomials $(f_k(t_1, \cdots, t_r; x_1), g_k(t_1, \cdots, t_r))$, with $f_k(t_1, \cdots, t_r; x_1) \in Q[t_1, \cdots, t_r; x_1]$, $g_k(t_1, \cdots, t_r) \in Q[t_1, \cdots, t_r]$, $k = 1, 2, \cdots, s$, such that for each $t_i = t_{i0}$ in a real-closed field $K, i = 1, 2, \cdots, r$,*

$$f(t_{10}, \cdots, t_{r0}; x_1, x_2) = 0, \quad g(t_{10}, \cdots, t_{r0}; x_1) \neq 0 \tag{2.20}$$

is solvable for $x_1, x_2 \in K$ if and only if one of the following conditions is solvable for x_1 in $\mathbf{K}$:

$$f_k(t_{10}, \cdots, t_{r0}; x_1) = 0, \quad g_k(t_{10}, \cdots, t_{r0}) \neq 0. \tag{2.21}$$

A detailed discussion leading to the proof of this theorem is contained in [33, pp. 308–312] and the interested reader is referred to the excellent treatment given there. Theorem 2.2 leads to the result stated next, which in essence is a generalization of the classical theorem due to Sturm.

Theorem 2.3. *Consider any collection $\{C\}$ of polynomial equality/inequality sets involving signs, $=, >, <, \geq, \leq,$ and $\neq$, where each polynomial belongs to $Q[t_1, \cdots, t_r; x_1, \cdots, x_n]$ (again Q is the field of rational numbers). Then, it is possible to determine in a finite number of steps a finite collection of finite sets $\{C_j\}$ of polynomial equations and inequalities of the same type, in the parameters $t_1, \cdots, t_r$ alone, such that the collection $\{C\}$ has a solution in a real-closed field $\mathbf{K}$ for the indeterminates $x_1, x_2, \cdots, x_n$ with the parameters t_i set at, say, $t_i = t_{i0} \in \mathbf{K}$ for $i = 1, 2, \cdots, r$, if and only if the r-tuple $(t_{10}, \cdots, t_{r0})$ satisfies all the conditions of one of the sets $\{C_j\}$.*

Proof. From Fact 2.5, it is sufficient to consider the solvability in $\mathbf{K}$ of a polynomial equation of the type

$$f(t_1, t_2, \cdots, t_r; x_1, \cdots, x_n) = 0, \tag{2.22}$$

where $f(t_1, \cdots, t_r; x_1, \cdots, x_n) \in Q[t_1, \cdots, t_r; x_1, \cdots, x_n]$ and $t_i \in \mathbf{K}, i = 1, 2, \cdots, r$ are parameters. It will be shown via the principle of mathematical induction that there exist a finite number of polynomials $g_k(t_1, \cdots, t_r) \in Q[t_1, \cdots, t_r], k = 1, 2, \cdots, s$ and an equal number of polynomials $f_k(t_1, \cdots, t_r; x_1) \in Q[t_1, \cdots, t_r; x_1]$ such that for $t_i = t_{i0} \in \mathbf{K}, i = 1, 2, \cdots, r$, the equation in (2.18) with each t_i specialized to t_{i0} has a solution in $\mathbf{K}$ if and only, if for at least one k in $k = 1, 2, \cdots, s,$

$$f_k(t_{10}, \cdots, t_{r0}; x_1) = 0, \quad g_k(t_{10}, \cdots, t_{r0}) \neq 0$$

has a solution in $\mathbf{K}$, where $f_k(t_1, \cdots, t_r; x_1), g_k(t_1, \cdots, t_r)$ are computable in a finite number of steps via rational operations.

The preceding statement is trivially true for $n = 0, n = 1$, and the truth of the statement for the $n = 2$ case follows from Theorem 2.2. Assume that the statement is valid for $n - 1$ variables $x_i, i = 1, 2, \cdots, n - 1$, where $n > 2$. With x_1 treated as a parameter, polynomials $f_{k1}(t_1, \cdots, t_r; x_1, x_2), g_{k1}(t_1, \cdots, t_r; x_1), k = 1, 2, \cdots, s_1,$ can be computed such that for any specialization, $t_i = t_{i0} \in \mathbf{K}, i = 1, 2, \cdots, r, x_1 = x_{10} \in \mathbf{K}$, the equation

$$f(t_{10}, \cdots, t_{r0}; x_{10}, x_2, \cdots, x_n) = 0$$

has a solution in $\mathbf{K}$ if and only if at least one of the equality/inequality

$$f_{k1}(t_{10}, \cdots, t_{r0}; x_{10}, x_2) = 0, \quad g_{k1}(t_{10}, \cdots, t_{r0}; x_{10}) \neq 0,$$

for $k = 1, 2, \cdots, s_1$ has a solution in $\mathbf{K}$. It then follows that for any $t_i = t_{i0} \in \mathbf{K}, i = 1, 2, \cdots, r$, (2.18) with each t_i specialized to t_{i0} has a solution in $\mathbf{K}$ if and only if, for some k in $k = 1, 2, \cdots, s_1,$

$$f_{k1}(t_{10}, \cdots, t_{r0}; x_1, x_2) = 0, \quad g_{k1}(t_{10}, \cdots, t_{r0}; x_1) \neq 0 \tag{2.23}$$

is solvable in $\mathbf{K}$. Direct use of Theorem 2.2 then leads to a finite set of polynomials $f_{k1j}(t_1, \cdots, t_r; x_1), g_{k1j}(t_1, \cdots, t_r), j = 1, 2, \cdots, j_k$, obtainable in a finite number of steps via rational operations for each k in $k = 1, 2, \cdots, s_1$ in (2.10), such that the equality/ inequality set in (2.10), has a solution in $\mathbf{K}$ if and only if at least one of the equality/ inequality sets of the form,

$$f_{k1j}(t_{10}, \cdots, t_{r0}; x_1) = 0, \quad g_{k1j}(t_{10}, \cdots, t_{r0}) \neq 0 \qquad (2.24)$$

has a solution of $\mathbf{K}$, where $j = 1, 2, \cdots, j_k$. Solvability in $\mathbf{K}$ of (2.20) can be decided via rational operations in a finite number of steps by making use of the classical theorem due to Sturm. The proof of the theorem is now complete.

It has been shown that if r is the number of multivariate polynomials occurring in a prenex input formula and if m is the maximum degree of any such polynomial in any variable, then the maximum computing time required to implement Seidenberg's decision procedure is, like Tarski's scheme, exponential in both r and m, for a fixed n.

2.2.4 Collins' Procedure

This quantifier elimination algorithm accepts as input any standard prenex formula of the form

$$(Q_1 x_1)(Q_2 x_2) \cdots (Q_i x_i) P(x_1, x_2, \cdots, x_n) \qquad (2.25)$$

(where $P(x_1, \cdots, x_n)$ is a quantifier-free standard formula constructed from atomic formulas involving signs, $=, >, <, \neq, \geq,$ and, $\leq,$ and each $(Q_i x_i)$ is either an existential or universal quantifier) and as output produces an equivalent standard quantifier-free formula $T(x_{i+1}, \cdots, x_n)$. It has been shown that the maximum computing time of the procedure is dominated by

$$(2m)^{2^{(2n+8)}} r^{2^{(n+6)}} l^3 s,$$

where m is the maximum degree of any polynomial (in $J[x_1, \cdots, x_n]$, where J is any commutative ring with identity, e.g., the ring of integers) in any variable x_i in the prenex input formula, n is the number of variables, r is the number of polynomials occurring in the input formula, l is the maximum length of any coefficient belonging to J in the formula, and s is the number of atomic formulas from which the input prenex formula is constructed. If $\{p_1, \cdots, p_r\}$ is the set of all polynomials occurring in (2.25), the procedure establishes that there is a decomposition of the n-dimensional real space R^n, into a finite number of disjoint connected sets called cells, in each of which each polynomial in the set $\{p_1, \cdots, p_r\}$ is sign invariant. These cells are cylindrically arranged with respect to each of the n variables, and their boundaries are the zeros of certain polynomials derivable in a finite number of steps from the polynomials in $\{p_1, \cdots, p_r\}$. These polynomials, whose zeros determine the boundaries of the cells, are the result of successive projections of a set of polynomials in $k-1$ indeterminates, for $k = n, n-1, \cdots, 2$. The cylindrical arrangement of cells is ensured by a condition called delineability of roots, defined next.

2.3 Sum of Squares (SOS) Representation and Robust Optimization

There has been a surge of interest during the last decade in the topics of semidefinite programming, semialgebraic sets (defined by multivariate polynomial equations, inequations and inequalities), robust optimization and sum-of-squares representation of classes of nonnegative definite multivariate polynomials and forms for applications in analysis and synthesis of control systems.

Definition 2.11. A basic semialgebraic set is a subset of $\mathbf{R}^n$ defined by a finite number of polynomial equations and inequalities.

Example 2.2. (1)

$$\left\{ (x_1, x_2) \in \mathbb{R}^2 \,\bigg|\, \frac{x_1^2}{3^2} + \frac{x_2^2}{2^2} \leq 1, x_1^2 - x_2 \leq 0 \right\}$$

(2)

$$f(\mathbf{x}) \in \mathbb{R}[\mathbf{x}], f(\mathbf{x}) > 0, \forall \mathbf{x}, \mathbf{x} \triangleq (x_1, \ldots, x_n)$$

Approaches for (2) include:

1. Elementary decision algebra methods (A. Tarski, Seidenberg, Collins, N. K. Bose, B. D. O. Anderson, E. I. Jury), already discussed in this chapter.
2. Gram matrix method (N. K. Bose, C. C. Li, M. D. Choi, T. Y. Lam, B. Reznik) [40].
3. Sum of squares (SOS) representation, when possible to do (SOSTOOLS, SEDUMI), because positive multivariate polynomials or forms may not always be representable as a sum of squares or forms.
4. Semidefinite programming to test feasibility of algebraic sets (C.N. Delzell (1980), P. A. Parrilo, B. Sturmfels) [41, 42].
5. Global lower bound approach (N. Z. Shor) [43].

2.3.1 SOS Decomposition by Gram Matrix Method

[40, 44–46].

Fact 2.9. *A multivariate real coefficient polynomial* $p(\mathbf{x})$ *in n real variables* $\mathbf{x} \triangleq$ $(x_1, \ldots, x_n)$ *and of total degree 2d is a SOS if and only if it is representable as* $p(\mathbf{x}) = \mathbf{v}^T Q \mathbf{v}$, *where the* $\binom{n+d}{d}$ *vector of monomials,*

$$\mathbf{v}^T = (1 \; x_1 \; x_2 \; \ldots \; x_n \; x_1 x_2 \; \ldots \; \ldots \; x_n^d)$$

and Q is a symmetric PSD (positive semidefinite) matrix.

2.3.1.1 Nonnegativity of a Polynomial $f(\mathbf{x})$ on an Algebraic Variety

Let $h_i(\mathbf{x}) = 0$ be constraints and let $\mathbf{I}$ denote the polynomial ideal $\mathbf{I} = \langle h_1(\mathbf{x}), \ldots, h_l(\mathbf{x}) \rangle$. Then, there exist polynomials $\lambda_i(\mathbf{x}) \in \mathbb{R}[\mathbf{x}]$ such that $f(\mathbf{x}) + \sum_i \lambda_i(\mathbf{x}) h_i(\mathbf{x})$ is a SOS in n-variate polynomial ring $\mathbb{R}[\mathbf{x}]$ if and only if $f(\mathbf{x}) + \mathbf{I}$ is a SOS in the quotient ring $\mathbb{R}[\mathbf{x}]/\mathbf{I}$. Under these equivalent conditions, $f(\mathbf{x})$ is nonnegative on the real variety $\{\mathbf{x} \in \mathbb{R}^n | h_i(\mathbf{x}) = 0, \forall i\}$ [47, pp. 187–188].

Example 2.3. This example is considered in [47, pp. 187–188], where some errors that occur are corrected below.

Is $f(\mathbf{x}) = 10 - x_1^2 - x_2$ nonnegative on $x_1^2 + x_2^2 - 1 = 0$? $\mathbf{I} = \langle x_1^2 + x_2^2 - 1 \rangle$ in this case of one constraint equation, $h(\mathbf{x}) = x_1^2 + x_2^2 - 1$ is the Gröbner basis of the corresponding ideal

$$10 - x_1^2 - x_2 = (1 \; x_1 \; x_2) \begin{pmatrix} q_{11} & q_{12} & q_{13} \\ q_{12} & q_{22} & q_{23} \\ q_{13} & q_{23} & q_{33} \end{pmatrix} \begin{pmatrix} 1 \\ x_1 \\ x_2 \end{pmatrix}$$

$$= q_{11} + q_{22}x_1^2 + q_{33}x_2^2 + 2q_{12}x_1 + 2q_{13}x_2 + 2q_{23}x_1 x_2 \qquad (2.26)$$

In the quotient ring $\mathbb{R}[\mathbf{x}]/\mathbf{I}$,

$$f(\mathbf{x})(\bmod \mathbf{I}) = (q_{11} + q_{22}) + (q_{33} - q_{22})x_2^2 + 2q_{12}x_1 + 2q_{13}x_2 + 2q_{23}x_1 x_2$$

$$= 9 + x_2^2 - x_2 \triangleq \mathbf{v}^T Q_1 \mathbf{v}, \qquad (2.27)$$

where $\mathbf{v}$ denotes the vector in Fact 2.9 above corresponding to this example and

$$Q_1 = \begin{pmatrix} 9 & 0 & -1/2 \\ 0 & 0 & 0 \\ -1/2 & 0 & 1 \end{pmatrix} = L^T L, L = \begin{pmatrix} 3 & 0 & -1/6 \\ 0 & 0 & \sqrt{35}/6 \end{pmatrix}$$

$$\Rightarrow 10 - x_1^2 - x_2 \equiv \left(3 - \frac{x_2}{6}\right)^2 + \frac{35}{36}x_2^2 (\bmod \mathbf{I})$$

$$\Rightarrow f(x_1, x_2) \text{ is a SOS on } \mathbb{R}[x_1, x_2]/\mathbf{I}$$

Therefore, SOS on quotient ring $\mathbb{R}[\mathbf{x}]/\mathbf{I}$ is needed, where $\mathbf{I} = \langle h_i(\mathbf{x}) \rangle_{i=1}^{l}$ is the ideal generated by equality constraints. The computations can be effectively done in $\mathbb{R}[\mathbf{x}]/\mathbf{I}$ after computing the Gröbner basis for $\mathbf{I}$ [47].

2.3.2 Positivstellensatz

A concept central in real algebraic geometry (like Hilbert's Nullstellensatz in complex algebraic geometry) is stated next.

Theorem 2.4 (Stengle's Positivstellensatz (1974)). *Given polynomials*

$$\{f_1, \ldots, f_r\}, \{g_1, \ldots, g_k\} \text{ and } \{h_1, \ldots, h_l\} \text{ in } \mathbf{x} = (x_1, \ldots, x_n),$$

the following are equivalent:

1.

$$\left\{ \mathbf{x} \in \mathbb{R}^n \; \middle| \; \begin{array}{l} f_i(\mathbf{x}) \geq 0, i = 1, 2, \ldots, r \\ g_i(\mathbf{x}) \neq 0, i = 1, 2, \ldots, k \\ h_i(\mathbf{x}) = 0, i = 1, 2, \ldots, l \end{array} \right\}$$

is the empty set.

2. There exist polynomials $f \in$ (cone generated by $\{f_1, \ldots, f_r\}$), $g \in$ (cone generated by $\{g_1, \ldots, g_k\}$), and $h \in$ (cone generated by $\{h_1, \ldots, h_l\}$) such that $f + g^2 + h = 0$.

Comments

1. The multiplicative monoid M generated by $\{g_i\}_{i=1}^{k}$ is the set of all finite products of g_i's including 1. e.g.

$$M(g_1, g_2) = \{g_1^{k_1}, g_2^{k_2} | k_1, k_2 \in \mathbb{Z}_+ \cup \{0\}\}$$

2. The cone generated by $\{f_i\}_{i=1}^{r}$ is

$$P(f_1, \ldots, f_r) = \left\{ s_0 + \sum_{i=1}^{l} s_i b_i | l \in \mathbb{Z}_+, s_i \in \Sigma_n, b_i \in M(f_1, \ldots, f_r) \right\}$$

where Σ_n denotes the set of SOS polynomials in n-variables. Note that $f_i^2 s_i \in \Sigma_n$ as well.

3. Positivstellensatz gives a characterization of the *infeasibility* of polynomial equations and inequalities over the reals.

The Positivestellensatz (P-satz) is useful because it provides a characterization of the infeasibility (refutation) of a system of polynomial equations and inequalities in conjunction with polynomial SOS and is beginning to be used in control theory [42].

Chapter 3
Multidimensional Sampling

3.1 Introduction

Digital filtering is used to process discrete data, obtained either from sampling continuous signals or in some other manner. Its range of application is extensive, including the processing of geophysical, biomedical, television and video, sonar, and radar data. Most of the discussion in this chapter applies to multidimensional problems, though for brevity in exposition the 2-D problem will be emphasized. In cases where brevity is not sacrificed, the results will be stated in the general n-D format, and situations where the generalization from the $n = 2$ case cannot be made in a straightforward manner, will be identified. This philosophy will be adhered to not only in this chapter but also in the rest of the book.

Various strategies exist to sample band-limited multidimensional signals. Consistent with current practice, parentheses and square brackets are used, respectively, around continuous variables and discrete integer-valued indices. Rectangular sampling of a 2-D analog waveform $g_a(x_1, x_2)$ produces the discrete signal (k_1, k_2 are integer valued)

$$g[k_1, k_2] = g_a(k_1 X_1, k_2 X_2),$$

where X_1 and X_2 are the horizontal and vertical uniform sampling periods along each of the two orthogonal axes, generating an uniform orthogonal sampling raster. Let F and IF denote, respectively, the Fourier transform and inverse Fourier transform operators. The 2-D Fourier transform of $g_a(x_1, x_2)$ is

$$G_a(\Omega_1, \Omega_2) \triangleq F[g_a(x_1, x_2)]$$

$$= \int_{-\infty}^{\infty} \int_{-\infty}^{\infty} g_a(x_1, x_2) e^{-j(\Omega_1 x_1 + \Omega_2 x_2)} dx_1 dx_2.$$

© Springer International Publishing AG 2017
N.K. Bose, *Applied Multidimensional Systems Theory*,
DOI 10.1007/978-3-319-46825-9_3

It will be assumed that $g_a(x_1, x_2)$ is such that $G_a(\Omega_1, \Omega_2)$ exists. There are a number of sets of sufficient conditions on $g_a(x_1, x_2)$ each of which guarantees existence of the integral above, and the theory of multidimensional Fourier transformation both in its classical and in its distributional sense (generalized Fourier transform of generalized functions) is richly documented in the literature [48]. The inverse Fourier transform relation is

$$g_a(x_1, x_2) \triangleq IF[G_a(\Omega_1, \Omega_2)]$$
$$= \frac{1}{(2\pi)^2} \int_{-\infty}^{\infty} \int_{-\infty}^{\infty} G_a(\Omega_1, \Omega_2) e^{j(\Omega_1 x_1 + \Omega_2 x_2)} d\Omega_1 d\Omega_2.$$

If $g_a(x_1, x_2)$ is bandlimited to a band-region D^2, then

$$G_a(\Omega_1, \Omega_2) = 0, \qquad (\Omega_1, \Omega_2) \notin D^2,$$

and the support of $G_a(\Omega_1, \Omega_2)$ is then said to be D^2. Bandlimitedness implies square integrability of the signal spectrum $G_a(\Omega_1, \Omega_2)$. In the case of a rectangular band-region, D^2 is the Cartesian product of the open intervals $(-\Omega_{1c}, \Omega_{1c})$ and $(-\Omega_{2c}, \Omega_{2c})$ for some $\Omega_{1c} > 0$ and $\Omega_{2c} > 0$. In the case of a low-pass rectangular band-region, for $g_a(x_1, x_2)$ to be exactly recoverable from the array $\{g[k_1, k_2]\}$, the inequalities

$$X_1 \leq \pi/\Omega_{1c}, \qquad X_2 \leq \pi/\Omega_{2c}$$

must hold (equalities hold for the 2-D counterpart of the *Nyquist rate* for temporal signals and will be referred to as such). In [49, 50] it is shown that rectangular sampling is a special case of a more general sampling strategy by which a bandlimited waveform is sampled on a nonorthogonal sampling raster. A special case of this general strategy is discussed in [51], where a hexagonal sampling raster is the subject of concern because of its relevance in phased array antennas.

The 2-D Fourier transform of the discrete array (bisequence or, for brevity, sequence may be considered to be acceptable in place of array) $\{g[k_1, k_2]\}$ is defined for notational brevity in terms of $[\omega_1, \omega_2]^T \triangleq \boldsymbol{\omega}$ as

$$G(e^{-j\omega_1}, e^{-j\omega_2}) \triangleq G(\boldsymbol{\omega}) = \sum_{k_1=-\infty}^{\infty} \sum_{k_2=-\infty}^{\infty} x[k_1, k_2] e^{-j(k_1\omega_1 + k_2\omega_2)}.$$

$G(\boldsymbol{\omega})$ is obtained by evaluating the 2-D z-transform $G(z_1, z_2)$, to be defined later, of sequence $\{g[k_1, k_2]\}$ at $z_1 = e^{-j\omega_1}, z_2 = e^{-j\omega_2}$. It may be assumed that $g_a(x_1, x_2)$ is sampled so that no aliasing occurs, i.e., $g_a(x_1, x_2)$ must be exactly recoverable from the sampled bisequence $\{g[k_1, k_2]\}$.

Indeed, then it is possible to show that from the output bisequence of the filter, a continuous function having Fourier transform $G_a(\Omega_1, \Omega_2)$, may be constructed.

If $\{g[k_1, k_2]\}$ whose Fourier transform is $G(e^{-j\omega_1}, e^{-j\omega_2})$ is filtered by a linear shift-invariant (LSI) 2-D digital filter having wavenumber response $H(e^{-j\omega_1}, e^{-j\omega_2})$ (which is defined to be the 2-D Fourier transform of the unit impulse response $\{h[k_1, k_2]\}$ of the 2-D digital filter), then the output bisequence has the product $H(e^{-j\omega_1}, e^{-j\omega_2})G(e^{-j\omega_1}, e^{-j\omega_2})$ for its Fourier transform. The overall design problem of a multidimensional recursive digital filter involves the various phases of approximation, realization, stabilization and stability, design optimization, and error analysis. The stability problem will be discussed here.

3.2 Multidimensional Sampling

In one and multidimensional signal processing, a continuous-space/time signal is usually represented and processed by its discrete samples. For a bandlimited signal, the classical Whittaker-Shannon-Kotelnikov (WSK) sampling theorem provides an exact representation from its uniformly spaced samples with sampling rate higher than or equal to the Nyquist rate along each of the mutually orthogonal space/time axis. For the sake of brevity in exposition and notation, the multidimensional sampling strategies are illustrated by discussing, initially, the two-dimensional case. Rectangular sampling leading to the reconstruction formula (referred to as the WSK sampling theorem) for recovery of the analog signal from its discrete samples is a routine generalization of 1-D results [52]. Rectangular sampling is neither the most general nor the most efficient way to sample multidimensional signals. Hexagonal sampling, which is more efficient than rectangular sampling, takes after the rods and cones in the human eye that are arranged in a honeycomb (hexagonal) fashion. Grasp of steps leading to reconstruction of rectangularly sampled signals, is, however, indispensable in applications and for appreciation of difficulties in generalization to the non-rectangular and nonuniform sampling cases.

Unless mentioned otherwise, the signals may be assumed to belong to the Paley-Wiener space $B_2^{\Omega_c}$ of bandlimited functions whose Fourier transforms are square-integrable (L_p space, with $p = 2$), and for the derivation of the WSK theorem, the support of the Fourier transforms is in $\Omega_c \triangleq [-\Omega_{1c}, \Omega_{1c}] \times [-\Omega_{2c}, \Omega_{2c}]$. When the Fourier transform contains delta functions associated with harmonic signals, this assumption is violated; however, the WSK reconstruction formula holds by sampling at a rate higher than the Nyquist rate, though the sampling series may not exhibit pointwise convergence (convergence in the distributional sense is possible). The reconstruction formula then holds by letting the Fourier transform have compact support in the Cartesian product, $(-\Omega_{1c}, \Omega_{1c}) \times (-\Omega_{2c}, \Omega_{2c})$ of two open intervals.

Besides the rectangular sampling pattern, the other sampling pattern, commonly used, is the hexagonal sampling pattern. Hexagonal sampling, to be discussed also in this chapter, is synonymous, subject to a scaling factor, to descriptors *interlaced* or *quincunx*.

3.2.1   Rectangular Sampling

Consider a spatial bandlimited analog signal $g(x_1, x_2)$ related to its Fourier transform $G(f_1, f_2)$, where $\Omega_i = 2\pi f_i$ for $i = 1, 2$, by the following relations:

$$F[g(x_1, x_2)] \triangleq G(f_1, f_2)$$

$$= \int_{-\infty}^{\infty} \int_{-\infty}^{\infty} g(x_1, x_2) exp[-j2\pi(f_1 x_1 + f_2 x_2)] dx_1 dx_2.$$

$$g(x_1, x_2) = IF[G(f_1, f_2)]$$

$$= \int_{-\infty}^{\infty} \int_{-\infty}^{\infty} G(f_1, f_2) exp[j2\pi(f_1 x_1 + f_2 x_2)] df_1 df_2.$$

The following properties of the 2-D *analog delta functional* $\delta(x_1, x_2)$, which is actually a generalized function, are useful for deriving the reconstruction formula from a sufficient number of sampled values. These properties are easy to justify by using the theory of distributions for generalized functions and the interested reader might wish to consult [53, 54].

Fact 3.1. *The direct product of the functionals* $\delta(x_1)$ *and* $\delta(x_2)$ *is* $\delta(x_1, x_2)$.

Proof. Use the property of distributions that: The direct product, $f(x_1) \cdot g(x_2)$ of two distributions $f(x_1)$ and $g(x_2)$ is another distribution that can be defined with respect to testing function $\phi(x_1, x_2)$ by

$$< f(x_1) \cdot g(x_2), \phi(x_1, x_2) > \triangleq < f(x_1), < g(x_2), \phi(x_1, x_2) >>$$
$$< \delta(x_1) \cdot \delta(x_2), \phi(x_1, x_2) > = < \delta(x_1), < \delta(x_2), \phi(x_1, x_2) >>$$
$$= < \delta(x_1), \phi(x_1, 0) >$$
$$= \phi(0, 0)$$
$$= < \delta(x_1, x_2), \phi(x_1, x_2) >$$
$$\delta(x_1) \cdot \delta(x_2) = \delta(x_2) \cdot \delta(x_1) = \delta(x_1, x_2).$$

□

Property 1. This property follows from the fact that the direct product of the 1-D delta functional with itself is a 2-D delta functional [53, p. 116] and the scaling property of 1-D delta functional,

$$\delta(\alpha_1 x_1, \alpha_2 x_2) = \frac{1}{|\alpha_1 \alpha_2|} \delta(x_1, x_2),$$

where α_1, α_2 are constants.

Noting that the Fourier transform of the unit step function $u(x_1)$ is $\left(\pi \delta(x_1) + \frac{1}{jw_1} \right)$, the Fourier transform of the 2-D unit step $u(x_1, x_2) = u(x_1)u(x_2)$ is $\left(\pi \delta(w_1) + \frac{1}{jw_1} \right) \left(\pi \delta(w_2) + \frac{1}{jw_2} \right) = \pi^2 \delta(w_1, w_2) + \frac{\pi}{j} \left[\frac{\delta(w_1)}{w_2} + \frac{\delta(w_2)}{w_1} \right] - \frac{1}{w_1 w_2}$.

Property 2. This property follows from the fact that the support of the direct product of two distributions is the Cartesian product of their supports [53, p. 118], so that the *brush* functional on the left side may be expressed as in the right side of the equation below with the help of Property 1.

$$\text{comb}(\alpha_1 x_1)\text{comb}(\alpha_2 x_2) = \frac{1}{|\alpha_1 \alpha_2|} \sum_{k_1 = -\infty}^{\infty} \sum_{k_2 = -\infty}^{\infty} \delta \left(x_1 - \frac{k_1}{\alpha_1}, x_2 - \frac{k_2}{\alpha_2} \right),$$

where the *comb* functional is defined as

$$\text{comb}(x) \triangleq \sum_{k=-\infty}^{\infty} \delta(x - k).$$

The Fourier transform of comb(x) is

$$\sum_{k=-\infty}^{\infty} \delta(f - k) = 2\pi \sum_{k=-\infty}^{\infty} \delta(\omega - 2\pi k).$$

Property 3. This property follows from the 1-D counterpart of Property 1 and the fact that the Fourier transform of an equi-spaced train of delta functionals with spacing X_1 and unit weighting is another such train [54, pp. 67–68] with spacing $\frac{1}{X_1}$ and weighting $\frac{1}{X_1}$.

$$F \left[\text{comb} \left(\frac{x_1}{X_1} \right) \right] = X_1 \text{comb}(X_1 f_1)$$

Property 4. This property follows from the fact that the Fourier transform of a product separable function is the product of their Fourier transform and after setting $X_1 = 1$ in Property 3.

$$F[\text{comb}(x_1)\text{comb}(x_2)] = \text{comb}(f_1)\text{comb}(f_2)$$

The sampled function

$$g_s(x_1, x_2) = \text{comb} \left(\frac{x_1}{X_1} \right) \text{comb} \left(\frac{x_2}{X_2} \right) g(x_1, x_2) \tag{3.1}$$

generated after sampling $g(x_1, x_2)$ by multiplying it appropriately as shown above, where X_1 and X_2 are the positive-valued spatial sampling periods, has a Fourier transform,

$$G_s(f_1,f_2) = F\left[\text{comb}\left(\frac{x_1}{X_1}\right)\text{comb}\left(\frac{x_2}{X_2}\right)\right] * * G(f_1,f_2) \tag{3.2}$$

which is expressed as a 2-D convolution of the Fourier transforms of the sampling signal and the original signal. Using Properties 3 and 4, one gets

$$F\left[\text{comb}\left(\frac{x_1}{X_1}\right)\text{comb}\left(\frac{x_2}{X_2}\right)\right] = X_1 X_2 \text{comb}(X_1 f_1)\text{comb}(X_2 f_2) \tag{3.3}$$

Subsequently, using Property 2,

$$X_1 X_2 \text{comb}(X_1 f_1)\text{comb}(X_2 f_2) = \sum_{k_1=-\infty}^{\infty}\sum_{k_2=-\infty}^{\infty} \delta\left(f_1 - \frac{k_1}{X_1}, f_2 - \frac{k_2}{X_2}\right) \tag{3.4}$$

and from (3.2) and (3.3) via use of (3.4), the spectrum or Fourier transform of the sampled function is

$$G_s(f_1,f_2) = \sum_{k_1=-\infty}^{\infty}\sum_{k_2=-\infty}^{\infty} G\left(f_1 - \frac{k_1}{X_1}, f_2 - \frac{k_2}{X_2}\right) \tag{3.5}$$

Thus the spectrum of $g_s(x_1, x_2)$ is the replication of the spectrum of $g(x_1, x_2)$ at each point, $(\frac{k_1}{X_1}, \frac{k_2}{X_2})$, for all integer values of k_1 and k_2, in the (f_1, f_2)-plane. Suppose that the bandlimited function $g(x_1, x_2)$ has a Fourier transform $G(f_1, f_2)$ whose support is strictly inside the rectangle defined by

$$-F_i < f_i < F_i, \quad i = 1, 2 \tag{3.6}$$

This rectangle is the smallest rectangle that completely encloses the region of the wavenumber space (i.e. (Ω_1, Ω_2)-space, where $\Omega_i = 2\pi f_i, i = 1, 2$) where $G(f_1, f_2)$ is nonzero. Then perfect reconstruction of $g(x_1, x_2)$ is possible from the sampled function $g_s(x_1, x_2)$ provided the sampling intervals are no greater than

$$X_i = \frac{1}{2F_i}, i = 1, 2 \tag{3.7}$$

associated with the Nyquist rate. The reconstruction formula is obtained via analog low-pass filtering of $G_s(f_1, f_2)$. Define a function,

$$\text{rect}(x) = \begin{cases} 1, & |x| \leq \frac{1}{2} \\ 0, & \text{otherwise.} \end{cases} \tag{3.8}$$

Then,

$$F[\text{rect}(x)] = \text{sinc}(f) \triangleq \frac{\sin(\pi f)}{\pi f} \tag{3.9}$$

Imagine an analog filter with a transfer function,

$$H(f_1, f_2) = \text{rect}\left(\frac{f_1}{2F_1}\right) \text{rect}\left(\frac{f_2}{2F_2}\right) \tag{3.10}$$

Provided Eq. (3.7) is satisfied, then

$$G_s(f_1, f_2)\text{rect}\left(\frac{f_1}{2F_1}\right) \text{rect}\left(\frac{f_2}{2F_2}\right) = G(f_1, f_2) \tag{3.11}$$

Denote the inverse Fourier transform of $H(f_1, f_2)$ by $h(x_1, x_2)$. Then,

$$h(x_1, x_2) = 4F_1 F_2 \text{sinc}(2F_1 x_1)\text{sinc}(2F_2 x_2) \tag{3.12}$$

The space-domain counterpart of (3.11) can be expressed as a 2-D convolution:

$$\left[\text{comb}\left(\frac{x_1}{X_1}\right) \text{comb}\left(\frac{x_2}{X_2}\right) g(x_1, x_2)\right] ** h(x_1, x_2) = g(x_1, x_2) \tag{3.13}$$

The equation below can be verified to be true.

$$\prod_{i=1}^{2} \text{comb}\left(\frac{x_i}{X_i}\right) g(x_1, x_2)$$

$$= X_1 X_2 \sum_{k_1=-\infty}^{\infty} \sum_{k_2=-\infty}^{\infty} g(k_1 X_1, k_2 X_2)\delta(x_1 - k_1 X_1, x_2 - k_2 X_2) \tag{3.14}$$

Therefore, the reconstruction formula, obtained from use of (3.14) in (3.13) is

$$g(x_1, x_2) = 4F_1 F_2 X_1 X_2 \sum_{k_1=-\infty}^{\infty} \sum_{k_2=-\infty}^{\infty} g(k_1 X_1, k_2 X_2)$$

$$\prod_{i=1}^{2} \text{sinc}(2F_i(x_i - k_i X_i)) \tag{3.15}$$

By sampling at the Nyquist rate along each axis

$$\Omega_{ic} \triangleq 2\pi F_i = \frac{\pi}{X_i}, i = 1, 2$$

the reconstruction formula in (3.15) can be expressed in the more compact form,

$$g(x_1, x_2) = \sum_{k_1=-\infty}^{\infty} \sum_{k_2=-\infty}^{\infty} g\left(k_1 \frac{\pi}{\Omega_{1c}}, k_2 \frac{\pi}{\Omega_{2c}}\right) \prod_{i=1}^{2} \text{sinc}\left(\frac{\Omega_{ic}}{\pi}\left(x_i - k_i \frac{\pi}{\Omega_{ic}}\right)\right). \tag{3.16}$$

For alternate proofs of the WSK sampling result in (3.16), see Problems 1 and 2.

3.2.2   Arbitrary Periodic Sampling Rasters

The real-valued and nonsingular sampling matrix V is of order 2 in the 2-D case. V is formed from its linearly independent column vectors, $\mathbf{v}_1$ and $\mathbf{v}_2$ as

Definition 3.1. Let $\mathbf{k} = [k_1 \ k_2]^T$ be a vector of integers. The set of all sample points $V\mathbf{k}$ obtained from all integer linear combinations of the columns of the sampling matrix V is called the lattice generated by V and denoted by LAT (V).

$$V = [\mathbf{v}_1 | \mathbf{v}_2].$$

An analog signal $g_a(x_1, x_2) \triangleq g_a(\mathbf{x})$ (again in vector notation, $\mathbf{x} = [x_1 \ \ x_2]^T$), after sampling on a raster, becomes the discrete signal,

$$g[k_1, k_2] \triangleq g[\mathbf{k}] = g_a(V\mathbf{k}).$$

Let $g_a(\mathbf{x})$ be a bandlimited analog signal whose CFT is $G_a(\mathbf{\Omega})$. Then

$$g_a(\mathbf{x}) = \frac{1}{4\pi^2} \int_{-\infty}^{\infty} \int_{-\infty}^{\infty} G_a(\mathbf{\Omega}) e^{j\mathbf{\Omega}^T \mathbf{x}} \ d\mathbf{\Omega},$$

$$g_a(V\mathbf{k}) \triangleq g[\mathbf{k}] = \frac{1}{4\pi^2} \int_{-\infty}^{\infty} \int_{-\infty}^{\infty} G_a(\mathbf{\Omega}) e^{j\mathbf{\Omega}^T V\mathbf{k}} \ d\mathbf{\Omega}.$$

After defining $\boldsymbol{\omega} = V^T \mathbf{\Omega}$, the previous equation becomes,

$$g[\mathbf{k}] = \frac{1}{|\det V| 4\pi^2} \int_{-\infty}^{\infty} \int_{-\infty}^{\infty} G_a((V^T)^{-1} \boldsymbol{\omega}) e^{j\boldsymbol{\omega}^T \mathbf{k}} \ d\boldsymbol{\omega}.$$

The Fourier transform of the bisequence $g[\mathbf{k}] = g_a(V\mathbf{k})$ is

$$G(e^{-j\omega_1}, e^{-j\omega_2}) \triangleq G(\boldsymbol{\omega}) = \sum_{k_1} \sum_{k_2} g[\mathbf{k}] e^{-j\boldsymbol{\omega}^T \mathbf{k}}$$

On integrating over the (ω_1, ω_2)-plane as an infinite series over square areas, each of size $2\pi \times 2\pi$,

$$g[\mathbf{k}] = \frac{1}{4\pi^2} \int_{-\pi}^{\pi} \int_{-\pi}^{\pi} \frac{1}{|\det V|} \sum_{\ell_1} \sum_{\ell_2} G_a((V^T)^{-1}(\boldsymbol{\omega} - 2\pi\boldsymbol{\ell})) e^{j\boldsymbol{\omega}^T \mathbf{k}} e^{-j2\pi\boldsymbol{\ell}^T \mathbf{k}} d\boldsymbol{\omega}$$

where $\boldsymbol{\ell} = [\ell_1 \ \ell_2]^T$ is an integer-valued vector. Since $e^{-j2\pi\boldsymbol{\ell}^T \mathbf{k}} = 1$, therefore comparing the above equation with the inverse discrete Fourier transform

$$g[\mathbf{k}] = \frac{1}{4\pi^2} \int_{-\pi}^{\pi} \int_{-\pi}^{\pi} G(\boldsymbol{\omega}) e^{j\boldsymbol{\omega}^T \mathbf{k}} d\boldsymbol{\omega}$$

of $G(\omega)$, it follows that

$$G(\omega) = \frac{1}{|\det V|} \sum_{\ell_1} \sum_{\ell_2} G_a((V^T)^{-1}(\omega - 2\pi\boldsymbol{\ell}))$$

The sampling density or the number of samples per unit area is proportional to $1/|\det V|$. Define FPD (V) to be the fundamental parallelepiped, defined as

$$\text{FPD}(V) : \{V\mathbf{x} \mid \mathbf{x} = \begin{bmatrix} x_1 \\ x_2 \end{bmatrix}, 0 \leqslant x_1, x_2, < 1\}$$

Let LAT (V) denote the set of all sample points $V\mathbf{k}$, where k is an integer-valued vector. The Fourier transform of the sampled signal is

$$G(V^T\boldsymbol{\Omega}) = \frac{1}{|\det V|} \sum_{\mathbf{k}} \sum G_a(\boldsymbol{\Omega} - U\mathbf{k}) \tag{3.17}$$

where the reciprocal lattice or polar lattice or aliasing matrix U satisfies the constraint,

$$U^T V = 2\pi I_2 \tag{3.18}$$

with I_2 being the identity matrix of order 2.

The term on the left-hand side of (3.17) can be interpreted as a periodic extension of $G_a(\boldsymbol{\Omega})$ obtained from copies of this bandlimited spectrum of the analog signal by shifting its origin to the points of LAT (U). This periodic extension is reflected in the property,

$$G(V^T(\boldsymbol{\Omega} + U\mathbf{k})) = G(V^T\boldsymbol{\Omega} + 2\pi\mathbf{k}) = G(V^T\boldsymbol{\Omega}) \tag{3.19}$$

The last equality in the previous equation follows because the Fourier transform $G(\omega)$ of the bisequence $x[\mathbf{k}]$ is 2π-*periodic* in ω_1 as well as in ω_2, and $\mathbf{k} = [k_1\ k_2]^T$ is an integer-valued vector.

The sampling and aliasing matrices, $V = V_R$ and $U = U_R$, respectively, in the case of the rectangular sampling raster considered in this subsection are,

$$V_R = \begin{bmatrix} X_1 & 0 \\ 0 & X_2 \end{bmatrix}, \quad U_R = \begin{bmatrix} \frac{2\pi}{X_1} & 0 \\ 0 & \frac{2\pi}{X_2} \end{bmatrix}. \tag{3.20}$$

When the sampling matrix is a diagonal matrix, the type of sampling lattice characterizing the sampling process is called separable. The generalization of rectangular sampling to n-D is routine and the sampling matrix in that case is a diagonal matrix of order n.

Example 3.1. Let the sampling matrix be

$$V = \begin{bmatrix} 1 & -1 \\ 1 & 2 \end{bmatrix}$$

LAT (V) is partially shown by the filled dots in Fig. 3.1a. V is not unique and could be replaced by a matrix VE, where E is an integer-valued unimodular matrix, i.e. its determinant, $+1$ or -1, is a unit in the ring of integers.

Then, the aliasing matrix is

$$U = 2\pi(V^{-1})^T = \frac{2\pi}{3} \begin{bmatrix} 2 & -1 \\ 1 & 1 \end{bmatrix}$$

LAT (U) is shown partially in Fig. 3.1b by the filled dots.

$$\boldsymbol{\Omega} = \begin{bmatrix} \Omega_1 \\ \Omega_2 \end{bmatrix} = (V^{-1})^T \boldsymbol{\omega} = \frac{1}{3} \begin{bmatrix} 2 & -1 \\ 1 & 1 \end{bmatrix} \begin{bmatrix} \omega_1 \\ \omega_2 \end{bmatrix}$$

The spectrum of the bandlimited analog signal $g_a(\mathbf{x})$ is shown by the diamond-shaped region in Fig. 3.1c. The support of the Fourier transform of the sampled signal $G(\boldsymbol{\omega}) = G(V^T \boldsymbol{\Omega})$ is shown in Fig. 3.1d as a replication of $G_a(\boldsymbol{\Omega})$ at every lattice point of LAT (U), in the (Ω_1, Ω_2) plane. Finally, support of $G(\boldsymbol{\omega})$ in the (ω_1, ω_2) plane is shown in Fig. 3.1e and it is clear that low-pass filtering of the baseband, $-\pi \leqslant \omega_i \leqslant \pi, i = 1, 2$, will lead to the reconstruction of $g_a(\mathbf{x})$ from its samples at LAT (V) in Fig. 3.1a.

The hexagonal sampling process to be discussed next has a nonseparable sampling lattice.

3.2.3 Hexagonal Sampling

The anisotropy of the spatial angular frequency (wavenumber) response of the human visual system can be exploited by using a non-orthogonal sampling pattern with a reduced sampling density. Wavenumber responses with passbands in the shapes of a parallelogram, particularly a diamond, and hexagon are suitable for the sampling structure conversions of video signals.

For hyperspherical bandlimited functions in n-D wavenumber space, the problem of reconstruction from spatio-temporal samples over a sampling lattice having the smallest sampling density is linked to the geometrical problem of enclosing hyperspheres with a regular polytope, which along with its copies can tessellate the wavenumber space without overlaps and gaps. Hexagonal sampling provides the optimal sampling scheme for signals bandlimited over a circularly symmetric

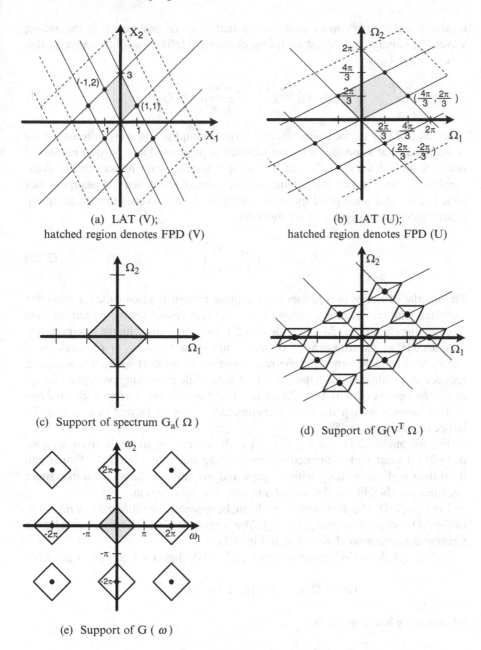

(a) LAT (V);
hatched region denotes FPD (V)

(b) LAT (U);
hatched region denotes FPD (U)

(c) Support of spectrum $G_a(\Omega)$

(d) Support of $G(V^T\Omega)$

(e) Support of $G(\omega)$

Fig. 3.1 (**a**) Sampling lattice LAT (*V*); (**b**) aliasing lattice LAT (*U*); (**c**) analog signal spectrum $G_a(\boldsymbol{\Omega})$; (**d**) sampled signal spectrum in (Ω_1, Ω_2); (**e**) sampled signal spectrum in (ω_1, ω_2)

region in wavenumber space in the sense that exact reconstruction of the analog waveform requires the lowest sampling density [51]. The sampling raster in this case is defined as

$$g[k_1, k_2] = g_a \left(\frac{2k_1 - k_2}{2} X_1, k_2 X_2 \right) \tag{3.21}$$

where X_1 and $2X_2$ are the horizontal and vertical sampling intervals. Alternate rows of the hexagonal sampling raster are identically positioned and the odd-numbered rows are shifted horizontally one-half sample interval with respect to the even-numbered rows. The sampling points are the corners and centers of hexagons that tessellate the spatio-temporal space. It is straightforward to verify that the sampling matrix V_H of the hexagonal sampling raster is

$$V_H = \begin{bmatrix} X_1 & X_1 \\ X_2 & -X_2 \end{bmatrix}. \tag{3.22}$$

The interlaced lattice of a hexagonal sampling pattern is often called a *quincunx* sampling pattern, because a group of five samples resembles the pattern of dots representing the number 5 on the side of a die (in conformity with the interpretation of quincunx as an arrangement of five symbols in a square or rectangle with one symbol at each corner and the remaining in the middle). Quincunx sampling reduces, in a simple manner, the amount of data while preserving perceptual image quality to a great extent. If $X_1 = 2X_2$, i. e. if the intervals between two horizontal and vertical samples are equal, then a quincunx lattice may be viewed as a rectangular lattice rotated through 45°.

For the analog 2-D signal $g_a(x_1, x_2)$ to be exactly recoverable from samples in (3.21) it must be bandlimited within a hexagonal band-region R. With R and its shifted replicas, a tiling without gaps and overlaps of the wavenumber space becomes possible, like in the case of a rectangular band-region.

Let $G_a(\Omega_1, \Omega_2)$ be the Fourier transform, hexagonally bandlimited to a region R, of the 2-D continuous signal $g_a(x_1, x_2)$. The support of $G_a(\Omega_1, \Omega_2)$ is shown by the hexagonal region around the origin in Fig. 3.2a.

$G(\Omega_1, \Omega_2)$, the periodic extension of $G_a(\Omega_1, \Omega_2)$, shown in Fig. 3.2b, is given by:

$$G(\Omega_1, \Omega_2) = G_a(\Omega_1, \Omega_2) * * D(\Omega_1, \Omega_2)$$

where for the hexagonal case

$$D(\Omega_1, \Omega_2) = \sum_{k_1=-\infty}^{\infty} \sum_{k_1=-\infty}^{\infty} \delta(\Omega_1 - k_1(2w_1 + w_3), \Omega_2 - 2k_2 w_2)$$

$$+ \sum_{k_1=-\infty}^{\infty} \sum_{k_2=-\infty}^{\infty} \delta \left(\Omega_1 - \left(k_1 + \frac{1}{2} \right) (2w_1 + w_3), \Omega_2 - (2k_2 + 1)w_2 \right)$$

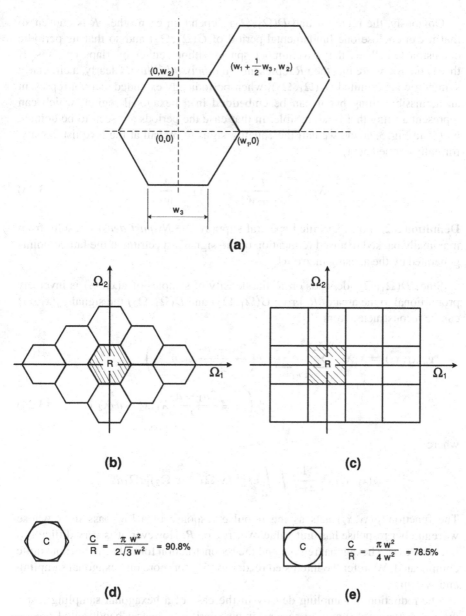

Fig. 3.2 Sampling densities for a circularly symmetric low-pass signal with rectangular and hexagonal tessellations in wavenumber space

Obviously, the region R and $D(\Omega_1, \Omega_2)$ depend on each other. R is chosen so that it can enclose one fundamental period of $G(\Omega_1, \Omega_2)$ and so that its periodic extension tessellates the wavenumber space without either overlaps or gaps. If these conditions are met then R represents an *admissible tiling*. Clearly, a circularly symmetric bandlimited $G_a(\Omega_1, \Omega_2)$, when periodically extended, cannot represent an admissible tiling but it can be embedded in a hexagonal region, which can represent a tiling that is admissible. In that case the periods are seen to be defined by (from Fig. 3.2a and we call the equality signs to prevail at the Nyquist density, formally defined next)

$$X_1 \leq \frac{4\pi}{2w_1 + w_3} \quad , \quad X_2 \leq \frac{\pi}{w_2}. \tag{3.23}$$

Definition 3.2. For a specified spectral support, the *Nyquist density* results from maximally packed unalised replication of the signal's spectrum at the lattice points generated by the aliasing matrix U.

Since $D(\Omega_1, \Omega_2)$ depends on R, the density of samples of $g(x_1, x_2)$ is inversely proportional to the area of R. From $G(\Omega_1, \Omega_2)$ and $D(\Omega_1, \Omega_2)$ the signal $g_a(x_1, x_2)$ can be reconstructed from

$$g_a(x_1, x_2) = X_1 X_2 \sum_{n_1 = -\infty}^{\infty} \sum_{n_2 = -\infty}^{\infty} g\left(\frac{2n_1 - n_2}{2} X_1, n_2 X_2\right)$$

$$\phi\left(x_1 - \frac{2n_1 - n_2}{2} X_1, x_2 - n_2 X_2\right) \tag{3.24}$$

where

$$\phi(x_1, x_2) = \frac{1}{4\pi^2} \int \int_R \exp[j(x_1 \Omega_1 + x_2 \Omega_2)] d\Omega_1 d\Omega_2.$$

The function $\phi(x_1, x_2)$ acts as the impulse response of a low-pass filter whose wavenumber response has unit value over region R. However, it is very difficult to integrate over a hexagonal region, and the reconstruction formula can become quite complicated. We refer the interested reader to [51] for more on hexagonal sampling and systems.

The reduction in sampling density in the case of a hexagonal sampling raster over a rectangular sampling raster for circularly symmetric bandlimited analog signals may be justified easily by considering the signal $g_a(x_1, x_2)$ to have a Fourier transform support R_G defined by

$$G_a(\Omega_1, \Omega_2) = \begin{cases} 1, & \Omega_1^2 + \Omega_2^2 < w^2, \quad (\Omega_1, \Omega_2) \in R_G \\ 0, & \Omega_1^2 + \Omega_2^2 \geq w^2, \quad \textit{otherwise.} \end{cases}$$

The support R_G can be enclosed within a square of side 2w and within a regular hexagon, each of whose six sides is of length $\frac{2w}{\sqrt{3}}$. From (3.20), it is clear that

$$|detV_R| = \frac{\pi^2}{w^2}.$$

From Fig. 3.2a, for the regular hexagon, $w_2 = w$, $w_1 = 3w$ and $w_3 = 2w/\sqrt{3}$, in this case. Therefore, at the Nyquist rate, from Eq. (3.23), the hexagonal sampling and aliasing matrices, $V = V_H$ and $U = U_H$, respectively, which satisfy the constraint in (3.18), are

$$V_H = \begin{bmatrix} \frac{\pi}{w\sqrt{3}} & \frac{\pi}{w\sqrt{3}} \\ \frac{\pi}{w} & -\frac{\pi}{w} \end{bmatrix} \quad , \quad U_H = \begin{bmatrix} w\sqrt{3} & w\sqrt{3} \\ w & -w \end{bmatrix}.$$

Therefore,

$$\frac{|\det V_H|}{|\det V_R|} = \frac{2}{\sqrt{3}}$$

The sampling density is proportional to $1/|\det V|$, when V is the sampling matrix. Thus by taking the ratio of $|\det V_H|$ and $|\det V_R|$, it can be seen that hexagonal sampling requires about 13.4 % less sample points to represent the same circularly symmetric bandlimited analog signal. In other words, for a specified sampling density, a hexagonal arrangement of samples can handle 13.4 % more bandwidth. Hexagons as well as rectangles can both provide admissible tiling in the sense that both can cover the wavenumber space without aliasing and holes as illustrated in Fig. 3.2b and 3.2c, respectively. Clearly, for a circularly symmetric bandlimited spectrum of a signal, enclosing regular hexagons can be more tightly packed than enclosing squares in an unit area where higher tightness is associated with a larger number of the basic blocks in that area. Since the area of a regular hexagon, each of whose sides is $2w/\sqrt{3}$, is smaller than the area of the square, each of whose sides is $2w$ (both embed a circle whose diameter is of the same length $2w$ as the side of the square) more hexagons than rectangles can be packed in the wavenumber space. This implies that a hexagon gives a tighter packing than a rectangle, or, equivalently the hexagonal raster sampling rate is lower than that of a rectangular raster. We conclude that hexagonal sampling requires less samples per unit area than rectangular sampling to completely represent a signal bandlimited over a circular region.

The efficiency of a sampling lattice depends on the area of support, R of the bandlimited signal, when R has the admissible tiling property. The only regular polygons that possess this property are equilateral triangles, squares and regular hexagons. Since the circular region R_G cannot tile the wavenumber space without holes and overlaps, the efficiency cannot be expressed only in terms of R_G. In that situation one finds an extended region R_E of area, say, R' with the admissible tiling property that also encloses R_G. Efficiency may then be defined by C/R' (where C is the area of the corresponding circular region R_G). The highest efficiency that a triangular scheme could achieve when the circularly bandlimited spectrum support is of radius r is

$$\frac{\pi r^2}{3\sqrt{3}r^2} = .604,$$

which is the lowest among the three regular polygons endowed with the admissible tiling property. Furthermore, tessellation of the wavenumber space with triangles require rotation in addition to shifting, a scheme which is difficult to implement. The corresponding efficiencies when the admissible tiling is done by a rectangular (actually, a square) and a regular hexagon are, respectively, 78.5 % and 90.8 % as shown in Fig. 3.2d and 3.2e.

The effect of sampling a spatiotemporal signal on a spatiotemporal lattice is to replicate the spectrum of the original signal on a reciprocal lattice in wavenumber space, as has been illustrated for rectangular and hexagonal lattices in the case of 2-D signals. It has been seen that to reconstruct the original signal from samples, the replicated copies of the spectrum should not overlap and cause aliasing. The theory of sampling multidimensional signals on lattice was presented by Petersen and Middleton [50] and exploited in [55] for video systems. The current relevance of video technology necessitate the inclusion, albeit brief, of multidimensional sampling within the framework of lattice and sublattice theory in the subsequent subsection.

Example 3.2. Consider the bandlimited 2-D spatial signal, whose support of the Fourier transform is shown in Fig. 3.3a.

To find the minimum sampling density for perfect reconstruction from uniformly-spaced samples when the signal is rectangularly sampled, enclose the wavenumber domain support in Fig. 3.3a in the manner shown in Fig. 3.3b. Then, the optimum sample spacing is

$$X_i = \frac{\pi}{4\pi} = \frac{1}{4} \text{ m/sample, for i} = 1, 2.$$

Thus, the sampling density associated with the Nyquist rate along each axis is

$$\frac{1}{X_1 X_2} = 16 \text{ samples/m}^2.$$

In case of hexagonal sampling, enclose the wavenumber domain support within a hexagon as shown in Fig. 3.3c. In this case,

$$X_1 = \frac{4\pi}{2w_1 + w_3} = \frac{4\pi}{12\pi + 4\pi} = \frac{1}{4}$$

$$X_2 = \frac{\pi}{w_2} = \frac{\pi}{4\pi} = \frac{1}{4}$$

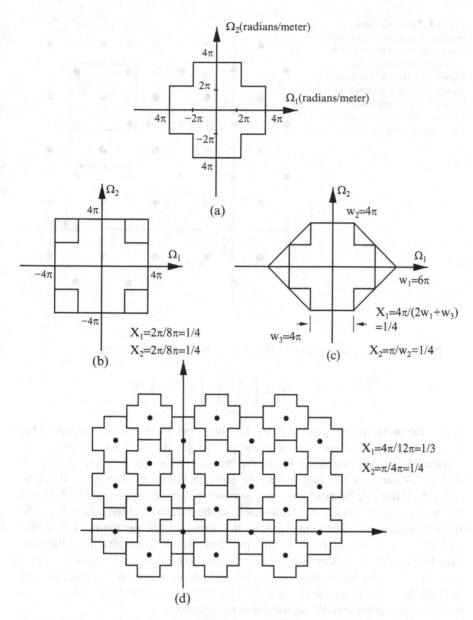

Fig. 3.3 Minimum sampling densities for different tessellations in Example 3.2

Fig. 3.4 The hexagonal, interlaced, quincunx sampling pattern. Note that around each sample are six other samples forming a hexagon. The basic quincunx pattern is shown by the five samples enclosed by the square in solid lines that result when $X_1 = 2 = 2X_2$

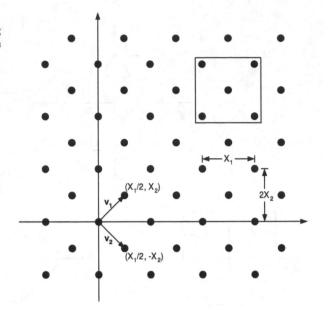

So, the sampling density given by $1/|\det V|$, is $16\,\text{samples/m}^2$, where

$$V = \begin{bmatrix} \frac{X_1}{2} & \frac{X_1}{2} \\ X_2 & -X_2 \end{bmatrix}$$

is the sampling matrix generated by the two vectors shown in Fig. 3.4. This illustrates that more than one sampling geometry may yield the sampling rate associated with that corresponding to the Nyquist rate along each axis. However, in 2-D this sampling rate may not be the absolute minimum as is justified below. Since the difference in area between wavenumber domain support in Fig. 3.3a and the enclosing square in Fig. 3.3b is the same as the difference of the area in Fig. 3.3a from the enclosing hexagon in Fig. 3.3c, it is expected that the sampling densities in both cases will be the same for this problem. To obtain the absolute minimum sampling density for this problem, it is necessary to enclose the hexagonal support R in Fig. 3.3a with the smallest region that will tile the plane. In this case, R itself will tile the plane as shown in Fig. 3.3d. The impulse train $D(\Omega_1, \Omega_2)$ that creates this periodic repetition of the signal Fourier transform is

$$D(\Omega_1, \Omega_2) = \sum_{k_1=-\infty}^{\infty} \sum_{k_1=-\infty}^{\infty} \delta(\Omega_1 - 12\pi k_1, \Omega_2 - 8\pi k_2)$$

$$+ \sum_{k_1=-\infty}^{\infty} \sum_{k_2=-\infty}^{\infty} \delta\left(\Omega_1 - \left(k_1 + \frac{1}{2}\right) 12\pi, \Omega_2 - (2k_2 + 1)4\pi\right)$$

The inverse Fourier transform of $D(\Omega_1, \Omega_2)$ can easily be verified to be

$$d(x_1, x_2) = \frac{X_1 X_2}{8\pi^2} \sum_{k_1} \sum_{k_2} [1 + (-1)^{k_1+k_2}] \delta\left(x_1 - \frac{k_1 X_1}{2}, x_2 - k_2\right)$$

This can be thought of as having two terms: one corresponding to the one term in the square bracket and another corresponding to the $(-1)^{k_1+k_2}$ term in the same square bracket. The resulting sampling raster is shown by the filled dots in Fig. 3.3d.

From above it is clear that

$$X_1 = \frac{2\pi}{6\pi} = \frac{1}{3} \quad , \quad X_2 = \frac{2\pi}{8\pi} = \frac{1}{4}$$

which provides the absolute minimum sampling density of 12 samples/m^2.

This minimum sampling density is obtained as $1/|\det V|$, where

$$V = \begin{bmatrix} \frac{X_1}{2} & \frac{X_1}{2} \\ X_2 & -X_2 \end{bmatrix} = \begin{bmatrix} \frac{1}{6} & \frac{1}{6} \\ \frac{1}{4} & -\frac{1}{4} \end{bmatrix}$$

so that $|\det V| = X_1 X_2 = \frac{1}{12}$.

An alternate way to find the minimum sampling density is to write down the U matrix from Fig. 3.3d. To wit,

$$U = \begin{bmatrix} 6\pi & 0 \\ 4\pi & 8\pi \end{bmatrix}$$

Then, a sampling matrix (non-unique) V_1 can be found from the constraint

$$V_1^T U = 2\pi I_2$$

$$V_1^T = \frac{2\pi}{48\pi^2} \begin{bmatrix} 8\pi & 0 \\ -4\pi & 6\pi \end{bmatrix} = \begin{bmatrix} \frac{1}{3} & 0 \\ -\frac{1}{6} & \frac{1}{4} \end{bmatrix}$$

Note, that $V_1 \neq V$ but $|\det V_1| = |\det V| = 1/12$. So, the minimum sampling density = 12.

3.2.4 n-D Sampling

In analog television the sequence of frames are sampled in vertical (spatial) and temporal directions using horizontal scanning with either an interlaced or progressive scanning raster. Two of the three spatiotemporal dimensions are discretized and the samples are not quantized in amplitude. In digital television, the horizontal spatial dimension is also sampled leading to a 3-D spatiotemporal discrete signal with a 3-D spectrum and with three-component discrete values for representation of color in an appropriate color space. Processing and coding of video signals require conversion between different sampling lattices used in different television standards and multidimensional filtering for picture quality improvement among other multidimensional analysis and synthesis approaches. The sampling process in n-D can be represented as a lattice defined as the set of all linear combinations over integers of n basis vectors $\mathbf{v}_1, \mathbf{v}_2, \ldots, \mathbf{v}_n$ (each of whose elements is an integer), in the matrix V that characterizes the sampling process.

$$V = [\mathbf{v}_1 \ \mathbf{v}_2 \ \ldots \ \mathbf{v}_n]$$

In certain applications the elements of V belong to the ring of integers and, therefore, the set of unimodular matrices over this ring has for its elements the matrices whose determinants are ± 1. This occurs, for example, when extending multiresolution analysis from 1-D to n-D [56]. A specific sampling pattern may not be represented by an unique matrix V. For a specified V the corresponding lattice, *Lat*, is the set of all vectors generated by

$$Lat \stackrel{\triangle}{=} V\mathbf{k}, \quad \mathbf{k} \in \mathcal{Z}^n.$$

An input cell is comprised of a set of points such that the disjoint union of its copies shifted to all lattice points yields the input lattice, which may be downsampled (decimated) or upsampled (interpolated). The magnitude of the determinant of V, $|detV|$, is the volume of the unit cell, which in turn is the reciprocal of the sampling density. The reciprocal lattice is the Fourier transform (spectrum) of the original lattice and its points represent locations at which replication of the spectrum occurs in the wavenumber space. The matrix, U, characterizing the reciprocal lattice satisfies the constraint

$$U^T V = (2\pi)I_n.$$

In the case of a downsampler if $x[\mathbf{k}]$ denotes the input n-D sequence, then the output n-D sequence

$$y[\mathbf{k}] = x[V\mathbf{k}],$$

has for its Fourier and z-transforms, the following expressions

$$Y(\omega) = \frac{1}{|detV|} \sum_{\mathbf{k} \in V_c^T} \cdots \sum X((V^T)^{-1}(\omega - 2\pi\mathbf{k})) \qquad (3.25)$$

$$Y(\mathbf{z}) = \frac{1}{|detV|} \sum_{\mathbf{k} \in V_c^T} \cdots \sum X(W_{V^{-1}}(2\pi\mathbf{k}) \circ \mathbf{z}^{D^{-1}}) \qquad (3.26)$$

3.2.4.1 Multidimensional Digital Analog Transformation

Analogue to the 1-D case discussed in the text [4], the multidimensional digital/analog (analog/digital) transformation can be efficiently implemented by the Q-matrix method. We focus on the trivariate case and the trend in the general case in obvious. Consider the trivariate polynomial,

$$B(z_1, z_2, z_3) = \sum_{k_1=0}^{N_1} \sum_{k_2=0}^{N_2} \sum_{k_3=0}^{N_3} a_k z_1^{k_1} z_2^{k_2} z_3^{k_3},$$

$$k \triangleq (N_1 - k_1)(N_2 + 1)(N_3 + 1) + (N_2 - k_2)(N_3 + 1) + (N_3 - k_3) + 1.$$

substitute $z_i = \frac{s_i+1}{s_i-1}$, $i = 1, 2, 3$, in $B(z_1, z_2, z_3)$ and the objective is to compute the numerator of the rational function polynomial $D(z_1, z_2, z_3)$ resulting from the transformation,

$$D(z_1, z_2, z_3) = \sum_{k_1=0}^{N_1} \sum_{k_2=0}^{N_2} \sum_{k_3=0}^{N_3} a_k \left(\prod_{i=1}^{3} (s_i - 1)^{N_i - k_i} (s_i + 1)^{k_i} \right).$$

First, we compute the term in the parenthesis in the preceding expression for each 3-tuple (k_1, k_2, k_3) associated with the relevant k. The mechanics of the procedure can best be illustrated by the example considered by Bose and Jury [57] using their original approach. Their example considers the trivariate polynomial to be written in recursive canonical form, with z_1 taken to be the main variable, as

$$B(z_1, z_2, z_3) = (4z_2 z_3 + z_2 - z_3) z_1^2 + z_2 z_3 z_1 + z_3.$$

Then, the coefficients in the above expression are written in recursive canonical form with z_2 as the main variable and, subsequently the nested form for $B(z_1, z_2, z_3)$ emerges,

$$B(z_1, z_2, z_3) = ((4z_3 + 1) z_2 - z_3) z_1^2 + z_3 (z_2)(z_1) + z_3.$$

Here, the partial degrees in the three variables z_1, z_2, z_3 are $N_1 = 2$, $N_2 = 1$, $N_3 = 1$. Consider the 3-tuple (k_1, k_2, k_3) to be set, quiet as $(2,1,1)$ so that $k = 1$ and the quiet column of the transformation matrix Q will be obtained from the coefficients of the polynomial product $(s_1 + 1)^2 (s_2 + 1)(s_3 + 1)$. The coefficients can be computed efficiently by convolution of powers involving same variable factors and Kronecker product across variables. To wit, the calculation involves the computation of

$$\{1, 1\} * \{1, 1\} \otimes \{1, 1\} \otimes \{1, 1\} = \{1, 1, 1, 1, 2, 2, 2, 2, 1, 1, 1, 1\}.$$

Including the zero-coefficient monomials, there are twelve monomials in $B(z_1, z_2, z_3)$, which can be arranged by a lexicographical ordering with $z_1 > z_2 > z_3$

as $z_1 z_2 z_3$, $z_1^2 z_2$, $z_1^2 z_3$, z_1^2, $z_1 z_2 z_3$, $z_1 z_2$, $z_1 z_3$, z_1, $z_2 z_3$, z_2, z_3, 1. Therefore, the remaining 11 columns of the 12×12 Q matrix can be computed similar to the first column and the coefficient vector of $D(s_1, s_2, s_3)$, after the same lexicographical ordering of the monomials, become $\mathbf{d} = Q\mathbf{b}$, where the coefficient vector,

$$\mathbf{b} = [4\ 1\ -1\ 0\ 1\ 0\ 0\ 0\ 0\ 0\ 1\ 0]^T$$

Note that the lacunary nature of $B(z_1, z_2, z_3)$ induces sparsity in the $\mathbf{b}$ vector as a result of which only the first three columns, the fifth column, and the eleventh column of Q need only be calculated.

3.3   Conclusions

The WSK sampling theorem generalizes to multidimensions in several ways. The most straightforward and routine generalization involves uniform sampling along each of the spatio-temporal dimensions when the function $g(x_1, x_2, \ldots, x_n)$ is bandlimited to the hyper-rectangle defined by the Cartesian product of the intervals $(-\Omega_{ic}, \Omega_{ic})$, $i = 1, 2, \ldots, n$. The possibility of sampling along nonorthogonal directions opens up possibilities for other types of generalizations. The sampling and reconstruction strategies involve the sampling matrix $\mathbf{V}$ and the aliasing matrix $\mathbf{U}$, which depend on each other in accordance with the constraint equation

$$\mathbf{U}^T \mathbf{V} = 2\pi \mathbf{I_n},$$

whose 2-D counterpart was seen in (3.18). The more general sampling rasters are considered with emphasis on hexagonal sampling. The exposition clarifies the relationship of minimum sampling density and tessellation of the wavenumber space without overlap and gaps (admissible tiling) by using a generic tile or period cell that encloses the spectrum of the analog bandlimited signal and its copies. The geometry of the sampling lattice depends on the tile shape (admissible tiling being only possible for only special-shaped period cells). In multidimensions (n-D) the sampling geometry corresponding to the sampling density (associated with Nyquist rate) for reconstruction without aliasing may be nonunique and also not necessarily minimum for $n > 1$.

The WSK sampling theorem has been generalized along various other directions. The reconstruction of a multivariate bandlimited function from irregularly spaced sample points has been a subject of recent research [58]. The reconstruction technique uses the fact that bandlimited functions satisfy a reproducing formula based on the convolution operation as in the WSK result. A sampling theorem for wavelet subspaces has been given for the 1-D case in [59]. There the multiresolution analysis of $L_2(\mathcal{R})$ corresponds to the subspaces $B_2^{2^m \pi}$, $m \in \mathcal{Z}$, the set of integers. A straightforward n-D generalization would be the building of a multiresolution analysis and wavelet bases in $L_2(\mathcal{R}^n)$ as the tensor product of a multiresolution analysis in $L_2(\mathcal{R})$. In $L_2(\mathcal{R}^2)$, it leads to a ladder of subspaces that are direct products of the ladder of closed subspaces in the 1-D case and then product separable wavelets are necessary to complete the analysis as in the 1-D case. However, a more general way of extending the sampling theorem to n-D involves the use of a dilation matrix with integer-valued entries each of whose singular values has a magnitude greater than unity.

Most of the multidimensional sampling algorithms rely on results from the bandlimited case, which may lead to unnecessarily high computational load, particularly for those classes of signals that could be represented by a finite number of samples. Sampling schemes and reconstruction formulae for certain classes of non-bandlimited multidimensional signals are now being developed.

3.4 Problems

1. For a square-integrable function, $g(x) \in L^2(-\infty, \infty)$, in the real variable x and having Fourier transform $G(\Omega)$ bandlimited to the interval defined by $-\Omega_c < \Omega < \Omega_c$, the following reproducing property is known to hold for an arbitrary but fixed value x_0 of the variable x.

$$\frac{\Omega_c}{\pi} \int_{-\infty}^{\infty} g(x) \frac{sin(\Omega_c(x - x_0))}{\Omega_c(x - x_0)} dx = g(x_0)$$

Define,

$$\phi_{mn}(x_1, x_2) \triangleq Sn(\Omega_1(x_1 - \frac{m\pi}{\Omega_1}))Sn(\Omega_2(x_2 - \frac{n\pi}{\Omega_2}))$$

where $Sn(x) \triangleq \frac{sin\ x}{x}$.

(a) Show that

$$I \triangleq \int_{-\infty}^{\infty} \int_{-\infty}^{\infty} \phi_{mn}(x_1, x_2)\phi_{kl}(x_1, x_2)dx_1 dx_2 = \frac{\pi^2}{\Omega_1 \Omega_2}\delta_{km}\delta_{ln}$$

where δ_{km} is the Kronecker delta function which is unity-valued when $k = m$ and zero otherwise.

(b) Now let $g(x_1, x_2) \in L^2(-\infty, \infty) \times L^2(-\infty, \infty)$. Suppose that the Fourier transform $G(\Omega_1, \Omega_2)$ of $g(x_1, x_2)$ is bandlimited so that the support of $G(\Omega_1, \Omega_2)$ is

$$Supp\ (G(\Omega_1, \Omega_2)) = \{(\Omega_1, \Omega_2) \mid -\Omega_{1c} < \Omega_1 < \Omega_{1c}, -\Omega_{2c} < \Omega_2 < \Omega_{2c}\}$$

Expand $g(x_1, x_2)$ in series

$$g(x_1, x_2) = \sum_{m=-\infty}^{\infty} \sum_{n=-\infty}^{\infty} c_{mn}\phi_{mn}(x_1, x_2)$$

Show that,

$$c_{mn} = \frac{\Omega_{10}\Omega_{20}}{\pi^2} \int\limits_{-\infty}^{\infty} \int\limits_{-\infty}^{\infty} g(x_1, x_2)\phi_{mn}(x_1, x_2)dx_1 dx_2$$

(c) Calculate the double integral in part (b) as an iterated operation using the reproducing property and arrive at the reconstruction formula in (3.15) for the case of rectangular sampling.

2. Prove the reconstruction result in (3.15) by applying the steps indicated below.

(a) For a fixed $x_2 = x_{20}$, expand $g(x_1, x_2)$ as a function of x_1, since it is bandlimited to the interval $-\Omega_{1c} < \Omega < \Omega_{1c}$, by the 1-D WSK sampling theorem.

$$g(x_1, x_{20}) = \sum_{k_1=-\infty}^{\infty} g(k_1\frac{\pi}{\Omega_{1c}}, x_{20})Sn(\Omega_{1c}(x_1 - k_1\frac{\pi}{\Omega_{1c}}))$$

(b) Similarly, expand as follows the function in the variable x_2,

$$g(k_1\frac{\pi}{\Omega_{1c}}, x_2) = \sum_{k_2=-\infty}^{\infty} g(k_1\frac{\pi}{\Omega_{1c}}, k_2\frac{\pi}{\Omega_{2c}})Sn(\Omega_{2c}(x_2 - k_2\frac{\pi}{\Omega_{2c}}))$$

after realizing that $g(x_1, x_2)$ as a function of x_2 for a fixed value of x_1 is bandlimited to the interval $-\Omega_{2c} < \Omega < \Omega_{2c}$.

(c) Combine the results in the previous two parts to get the result wanted.

3. Use the reproducing property in Problem 1 to arrive at its 2-D counterpart,

$$\int\limits_{-\infty}^{\infty} \int\limits_{-\infty}^{\infty} g(x_1, x_2)\frac{sin(\Omega_{1c}(x_1 - x_{10}))}{\pi(x_1 - x_{10})} \frac{sin(\Omega_{2c}(x_2 - x_{20}))}{\pi(x_2 - x_{20})} dx_1 dx_2 = g(x_{10}, x_{20})$$

when the square integrable bivariate function $g(x_1, x_2) \in L^2(-\infty, \infty)^2$ is bandlimited to the region defined by $-\Omega_{1c} < \Omega_1 < \Omega_{1c}, -\Omega_{2c} < \Omega_2 < \Omega_{2c}$.

(a) Show that the reconstruction result in (3.15) can be arrived at directly by applying the preceding reproducing property.

(b) Show that the following result holds

$$\sum_{k_1=0}^{\infty} \sum_{k_2=0}^{\infty} \frac{sin^2(k_1 a_1 + b_1)}{(k_1 a_1 + b_1)^2} \frac{sin^2(k_2 a_2 + b_2)}{(k_2 a_2 + b_2)^2} = \frac{\pi^2}{a_1 a_2}$$

for arbitrary but fixed real values of the parameters a_1, a_2, b_1, b_2 and $|a_1| \le \pi$, $|a_2| \le \pi$.

Chapter 4
Multidimensional Digital Filter Recursibility and Stability

4.1 Introduction

Given a continuous signal, acquired through sensors for subsequent sampling and discrete spatio-temporal processing, it may be assumed that, the sensed signal $g_a(x_1, x_2)$, in the 2-D case, is sampled so that no aliasing occurs, i.e., $g_a(x_1, x_2)$ must be exactly recoverable from the sampled bisequence $\{g[k_1, k_2]\}$. Indeed, then it was shown in the previous chapter, how from the output bisequence of the filter a continuous function having Fourier transform $G_a(\Omega_1, \Omega_2)$, may be constructed. If $\{g[k_1, k_2]\}$ whose Fourier transform is $G(e^{-j\omega_1}, e^{-j\omega_2})$ is filtered by a linear shift-invariant (LSI) 2-D discrete filter having wavenumber response $H(e^{-j\omega_1}, e^{-j\omega_2})$ (which is defined to be the 2-D Fourier transform of the unit impulse response $\{h[k_1, k_2]\}$ of the 2-D digital filter) then the output bisequence has the product $H(e^{-j\omega_1}, e^{-j\omega_2})G(e^{-j\omega_1}, e^{-j\omega_2})$ for its Fourier transform. The overall design problem of a multidimensional recursive digital filter involves the various phases of approximation, realization, stabilization and stability, design optimization, and error analysis. The stability problem will be discussed here first.

4.2 Recursible LSI Systems and Basic Tools for Their Stability Analysis

Stability properties in this chapter are studied with respect to linear shift-invariant systems, defined next.

© Springer International Publishing AG 2017
N.K. Bose, *Applied Multidimensional Systems Theory*,
DOI 10.1007/978-3-319-46825-9_4

Definition 4.1. A 2-D discrete system characterized by the operator $T[.]$ is said to be linear if and only if for arbitrary inputs $x_1[k_1, k_2], x_2[k_1, k_2]$ and any complex constants c_1, c_2,

$$T[c_1 x_1[k_1, k_2] + c_2 x_2[k_1, k_2]] = c_1 T[x_1[k_1, k_2]] + c_2 T[x_2[k_1, k_2]]. \tag{4.1}$$

If $y[k_1, k_2] = T[x[k_1, k_2]]$, the system characterized by $T[.]$ is shift-invariant if and only if

$$y[k_1 - k_0, k_2 - l_0] = T[x[k_1 - k_0, k_2 - l_0]] \tag{4.2}$$

for all $x[k_1, k_2]$, with k_0, l_0 arbitrary integers. The system satisfying both the above properties is linear shift-invariant (LSI).

The mathematical tool used in the study of multidimensional LSI systems is the multidimensional z-transform. In the case of 2-D systems, the z-transform of a sequence $\{x[k_1, k_2\}$ is defined to be:

$$Z[x[k_1, k_2]] \triangleq X(z_1, z_2) \triangleq \sum_{k_1 = -\infty}^{\infty} \sum_{k_2 = -\infty}^{\infty} x[k_1, k_2] z_1^{-k_1} z_2^{-k_2}. \tag{4.3}$$

An alternative definition replaces $z_1^{-k_1} z_2^{-k_2}$ with $w_1^{k_1} w_2^{k_2}$ in (4.3), but no essential conceptual difference arises due to these two possibilities. The expository advantages of working with the complex variables $w_1 \triangleq z_1^{-1}, w_2 \triangleq z_2^{-1}$ is context-dependent and this will be borne in mind in the text. In combinatorial studies, $X(z_1, z_2)$ is referred to as a generating function. The values of z_1, z_2 for which the double summation in (4.3) converges absolutely constitute the region of convergence, referred to as the *Reinhardt domain*. This domain is completely specified by the magnitudes $|z_1|, |z_2|$, respectively, of the complex variables z_1, z_2.

The inversion formula associated with (4.3) is:

$$x[k_1, k_2] = \frac{1}{(2\pi j)^2} \oint_{C_1} \oint_{C_2} X(z_1, z_2) z_1^{-1-k_1} z_2^{-1-k_2} dz_1 dz_2. \tag{4.4}$$

In (4.4), the closed contours C_1, C_2 must be in the region of convergence of $X(z_1, z_2)$ in (4.1). The Reinhardt domain must be specified before the sequence $x[k_1, k_2]$ may be uniquely calculated from its z-transform $X(z_1, z_2)$.

Any 2-D LSI system is completely specified by its impulse response

$$h[k_1, k_2] = T[\delta[k_1, k_2]],$$

where the 2-D unit impulse function is

$$\delta[k_1, k_2] = \begin{cases} 1, & \text{if } k_1 = k_2 = 0 \\ 0, & \text{otherwise} \end{cases}$$

The output $y[k_1, k_2]$ of any 2-D LSI system with impulse response sequence $\{h[k_1, k_2]\}$ and specified input $\{x[i_1, i_2]\}$ is given by the 2-D discrete convolution

$$y[k_1, k_2] = \sum_{i_1=-\infty}^{\infty} \sum_{i_2=-\infty}^{\infty} x[i_1, i_2] h[k_1 - i_1, k_2 - i_2]. \qquad (4.5)$$

The input-output relationship of a 2-D LSI system is defined by a difference equation of the form

$$\sum_{i_1} \sum_{i_2} b[i_1, i_2] y[k_1 - i_1, k_2 - i_2] = \sum_{i_1} \sum_{i_2} a[i_1, i_2] x[k_1 - i_1, k_2 - i_2]. \qquad (4.6)$$

$$(i_1, i_2) \in I_b \qquad\qquad\qquad (i_1, i_2) \in I_a$$

where I_b (output mask) and I_a (input mask) denote, respectively, the finite area regions of support for arrays $\{b[i_1, i_2]\}, \{a[i_1, i_2]\}$. With $b[0, 0] = 1$, (4.6) can be written as

$$y[k_1, k_2] = \sum_{i_1} \sum_{i_2} a[i_1, i_2] x[k_1 - i_1, k_2 - i_2] - \sum_{i_1} \sum_{i_2} b[i_1, i_2] y[k_1 - i_1, k_2 - i_2] \qquad (4.7)$$

$$(i_1, i_2) \in I_a \qquad\qquad\qquad (i_1, i_2) \in I_b$$
$$(i_1, i_2) \neq (0, 0)$$

Definition 4.2. Recursibiltiy (recursive computability) is defined to be a property of certain difference equations which allows one to iterate the equation by choosing an indexing scheme so that every output sample can be computed from outputs that have already been found, from initial conditions, and from samples of the input sequence. In that case, the system characterized by the difference equation is said to be recursively computable (recursive).

In discussing the recursibility of a 2-D difference equation, it is helpful to develop a generalization of the 1-D concept of the one-sided sequence. Recall that $\{h[n]\}$ is said to be right-sided with respect to some integer n_1 if

$$h[n] = 0 \quad \text{for } n < n_1$$

A left-sided sequence $h[n]$ with respect to some integer n_2 is one for which

$$h[n] = 0 \quad \text{for } n > n_2$$

A one-sided sequence is a sequence which is either right-sided or left-sided. Let us now introduce the concept of "one-sided" 2-D array.

Definition 4.3. Let H_1 be a line passing through the origin of the (k_1, k_2) plane. Let H_{1h} be a half-line starting at the origin (but not including the origin) and proceeding along H_1 in either of two possible ways; clearly, H_{1h} is contained in H_1. Let H_{2h} denote either of the two half-planes defined by the line H_1 but not including the line H_1. A 2-D array or bisequence $\{b[k_1, k_2]\}$ is one-sided if a line H_1, a half-line H_{1h}, and a half-plane H_{2h} can be found such that $b[k_1, k_2] = 0$ for all values of (k_1, k_2) lying on H_{1h} and H_{2h}. Such an array is said to be one-sided with respect to the line H_1.

An array $\{b[k_1, k_2]\}$ is one-sided if, for example,

$$b[k_1, k_2] = 0 \quad \text{for} \quad k_1 < 0 \quad \text{and for} \quad k_1 = 0, k_2 < 0$$

Note that, for the most part, the support of $\{b[k_1, k_2]\}$ is on "one side" of the ordinate where $k_1 = 0$. We will also consider the arrays generated by reflecting a one-sided array in either of both axes and rotating a one-sided array by 90, 180, or 270° to be one-sided. In this manner, seven additional examples of one-sided arrays are generated from the original $b[k_1, k_2]$. They are $b[k_1, -k_2], b[-k_1, k_2], b[-k_1, -k_2], b[k_2, k_1], b[k_2, -k_1], b[-k_2, k_1]$, and $b[-k_2, -k_1]$.

Consider next a line H_1 through the origin of the plane at any arbitrary angle θ with respect to the abscissa. An equation describing such a line is

$$k_1 \sin\theta - k_2 \cos\theta = 0 \tag{4.8}$$

The array $\{b[k_1, k_2]\}$ whose support excludes the region described below is one-sided with respect to the line generated by rotating the abscissa by an angle θ.

$$b[k_1, k_2] = 0 \quad \text{for} \quad \begin{aligned} k_1 \sin\theta - k_2 \cos\theta &< 0 \\ k_1 \sin\theta - k_2 \cos\theta &= 0 \\ k_1 \cos\theta + k_2 \sin\theta &< 0 \end{aligned}$$

The mirror-image set of the above one-sided arrays is generated by reflection with respect to the line in (4.8) and is described by the bisequence,

$$b[k_1, k_2] = 0 \quad \text{for} \quad \begin{aligned} k_1 \sin\theta - k_2 \cos\theta &> 0 \\ k_1 \sin\theta - k_2 \cos\theta &= 0 \\ k_1 \cos\theta + k_2 \sin\theta &> 0 \end{aligned}$$

These two sets of arrays whose supports are described in the preceding two equations may be considered analogous to the two sets (right-sided and left-sided) of sequences which are one-sided with respect to the origin. We are now ready to formalize the definition of one-sidedness.

Recursibility (also referred to as being recursible or having a recursible form) is defined as a property of certain 2-D difference equations which allows one to iterate the equation (that is, choose an indexing scheme) such that any point $y[k_1, k_2]$ may be calculated from a set of initial conditions and an input array.

Clearly, we must carefully pick the way we move the output mask around so as not to cover an output point which has not yet been calculated. Also, the input mask imposes no such constraint if we have the entire input before starting. Intuitively, if the hole in the output mask, which indicates the output point we are calculating at the moment, is in the middle or even on the edge of the mask, we cannot increment k_1 and k_2 in such a way as to be able to calculate $y[k_1, k_2]$ for all values of (k_1, k_2). Therefore difference equations with such output masks are not recursible. However, if the hole is on a corner of the mask, then there is no problem in iterating the difference equation.

Example 4.1. (a) Consider the difference equation

$$y[k_1, k_2] = x[k_1, k_2] - \sum_{l_1=1}^{2} \sum_{l_2=-2}^{2} a[l_1, l_2] y[k_1 - l_1, k_2 - l_2]$$

$$- \sum_{l_2=1}^{2} a[0, l_2] y[k_1, k_2 - l_2]$$

The input mask is very simple since its form is not important if the entire input is known beforehand; the output mask has a jog in it and the output hole is at the jog. Because of this, we may sweep the output mask along successive columns until we have generated all the output points of interest. The equation is recursible.

(b) Consider the difference equation

$$y[k_1, k_2] = x[k_1, k_2] - \frac{1}{2} y[k_1 - 1, k_2] - \frac{1}{4} y[k_1, k_2 - 1]$$

Here, $a[0, 0] = 1, a[1, 0] = \frac{1}{2}, a[0, 1] = \frac{1}{4}$; all other coefficients are zero. The array $\{b[k_1, k_2]\}$ is one-sided. Actually, the equation may be iterated in an infinite number of ways.

A recursively realizable representation based on the concept of one-sidedness is possible in the n-D ($n > 2$) case also. Heuristically, a n-D sequence $\{b[k_1, \cdots, k_n]\}$, where $\{(k_1, \cdots, k_n)\}$ is a subset of Z^n is one-sided with respect to the origin H_0 if there exists an $(n - 1)$-D hyperplane H_{n-1} passing through H_0 such that $\{b[k_1, \cdots, k_n]\}$ lies on one side of H_{n-1} inclusive, and within H_{n-1} there exists an $(n - 2)$-D hyperplane H_{n-2} such that $\{b[k_1, \cdots, k_n]\} \cap H_{n-1}$ lies on one side of H_{n-2} inclusive, etc., down to H_0. Again, the n-D counterpart of (4.7) is recursively realizable if and only if the weighting sequence $\{b[k_1, \cdots, k_n]\}$ associated with the output mask is one-sided. Though causality is not an intrinsic property of spatial

n-D systems,[1] a recursible n-D system can be described as causal (causality being interpreted as distinction between "past" and "future" in the spatial processing) with respect to some subset of the set of all orderings of Z^n.

4.3 Weak Causality and Recursibility

Eising [61, pp. 70] defined a *causality cone* C_c as an intersection of two half-plane, $H_{p,r}$ and $H_{q,t}$, where

$$H_{p,r} = \{(x_1, x_2) : px_1 + rx_2 \geq 0, (x_1, x_2) \in \mathbb{R}^2\}$$
$$H_{q,t} = \{(x_1, x_2) : qx_1 + tx_2 \geq 0, (x_1, x_2) \in \mathbb{R}^2\}$$

and p, q, r, t are nonnegative integers satisfying $pt - qr = 1$. Furthermore, Eising defined a 2-D filter to be *weakly causal* if the support of its impulse response $\{h[k_1, k_2]\}$ is contained within a *closed convex cone* C in $\mathbb{R}^2$ (i.e., $supp\{h[k_1, k_2]\} \subset C$), satisfying,

 (i) $C \cap (-C) = \{[0, 0]\}$ (this condition merely implies that the cone makes an angle of less than π at$[0, 0]$)
 (ii) $Q_1 \subset C$, where Q_1 denotes the first quadrant. Eising also showed that there exists a C_c such that for any weakly causal filter, $C \subset C_c$. For the sake of brevity we denote C_c by $H_{p,q,r,t}$, where p, q, r, t are integers defined above.

$H_{1,0,0,1}$ is the first quadrant, Q_1. A filter with support of its unit impulse response in Q_1 is called *causal*. A weakly causal filter can always be implemented recursively as will be justified here. It is important to note that if $\mathbb{Z}$ denote the set of integers and $\mathbb{Z}^2 \triangleq \mathbb{Z} \times \mathbb{Z}$ (Cartesian product of $\mathbb{Z}$ with itself) then there exists a map

$$\phi : (C_c \cap \mathbb{Z}^2) \triangleq S_{p,q,r,t} \rightarrow Q_1 \cap \mathbb{Z}^2 = S_{1,0,0,1}$$

given by

$$\phi[m, n] = [pm + rn, qm + tn], \qquad m, n \in S_{p,q,r,t}$$

which is bijective, because the matrix of transformation $\begin{bmatrix} p & r \\ q & t \end{bmatrix}$ is unimodular. $S_{p,q,r,t}$ is a semigroup under addition with $(0, 0)$ as "identity", which is generated by $(t, -q)$ and $(-r, p)$. The points $[t, -q]$ and $[-r, p]$ in C_c are mapped by ϕ to points $[1, 0]$ and

---

[1]Editorial note: The property of causality of n-D systems arising from modelling of physical phenomena is strictly dictated by the physics of the problem (and is thus an intrinsic property), the details of which have been recently elaborated in [60].

[0, 1], respectively, in Q_1. Suppose that the mapping

$$\phi : S_{p,q,r,t} \rightarrow S_{1,0,0,1}$$

is given by

$$\phi H(\alpha, \beta) = \sum_{k_1=0}^{\infty} \sum_{k_2=0}^{\infty} h\left[\phi^{-1}[k_1, k_2]\right] \alpha^{-k_1} \beta^{-k_2}$$

where $\phi[m, n]$ is as specified above. Then, it will be proved, after an example, that the isomorphism ϕ can be described by the substitutions

$$z_1 = \alpha^p \beta^q, \qquad z_2 = \alpha^r \beta^t$$

with inverse

$$\alpha = z_1^t z_2^{-q}, \qquad \beta = z_1^{-r} z_2^p$$

Example 4.2. Consider the transfer function of a digital filter

$$H(z_1, z_2) = \frac{Y(z_1, z_2)}{X(z_1, z_2)} = \frac{1}{1 + 0.5 z_1 z_2^{-1} + z_1^{-1} + z_2^{-1} + z_1^{-2} z_2}$$

It is stated that the filter is not of the quarter-plane type.[2] Cross-multiplication leads to

$$(1 + 0.5 z_1 z_2^{-1} + z_1^{-1} + z_2^{-1} + z_1^{-2} z_2) Y(z_1, z_2) = X(z_1, z_2)$$

The unit impulse response sequence $\{h[k_1, k_2]\}$ may be found if it is possible to implement the following difference equation (that follows from the preceding equation after replacing the input by the unit impulse), recursively, with zero boundary conditions.

$$h[k_1, k_2] = \delta[k_1, k_2] - \frac{1}{2} h[k_1 + 1, k_2 - 1] - h[k_1 - 1, k_2]$$
$$-h[k_1, k_2 - 1] - h[k_1 - 2, k_2 + 1],$$

where $\delta[k_1, k_2]$ represents the 2-D unit impulse. The output mask is obtained from the indices of the terms in the difference equation, as shown in Fig. 4.1.

---

[2]Editorial note: Quarter plane type filters are discussed later (cf. Fig. 4.3).

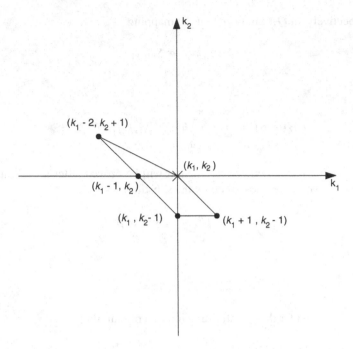

Fig. 4.1 Output mask; *cross* denotes present point and *filled dots* are associated with past points. Note that the output mask has an interior angle less than 180° at the corner containing the present point

In order to implement the desired recursion mark as coordinates in the $[k_1, k_2]$-plane the integer 2-tuples $[n_1, n_2]$ that can be associated with each monomial, $z_1^{-n_1} z_2^{-n_2}$, occurring in the denominator function of $H(z_1, z_2)$. A sector, where past, present and future points in the ordering, required to implement the recursion, are identified in Fig. 4.2.

The sector, bounded by lines OA ($x_1 + x_2 = 0$) and OB($x_1 + 2x_2 = 0$), extending from the origin $O = [0, 0]$ in the $[x_1, x_2]$-plane, which forms an angle of less than 180° at vertex O, contains the region of support of the filter impulse response. A causality cone, $H_{1,1,1,2}$ contains the support of this unit impulse response of the filter. $\{h[k_1, k_2]\}$ may be computed recursively as follows. First, obtain $h[k_1, k_2]$ on the boundaries OA and OB of the wedge (or sector or cone). Computation along OA is carried out recursively as

$$h[0, 0] = \delta[0, 0] - 0.5h[1, -1] - h[-1, 0] - h[0, -1] - h[-2, 1] = 1$$

$$h[-1, 1] = \delta[-1, 1] - 0.5h[0, 0] - h[-2, 1] - h[-1, 0] - h[-3, 2] = -0.5$$

$$h[-2, 2] = 0.25 \quad \dots\dots \quad h[-k, k] = (-0.5)^k, k \geqslant 0$$

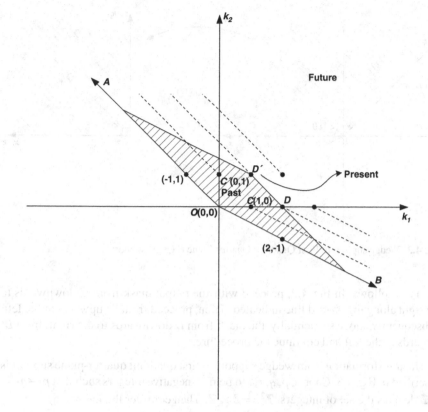

Fig. 4.2 Parallelogram containing past (recursively computed) points with the present point at one corner in the region of support of the unit impulse response. The region of support is only within the wedge created by the lines OA and OB (forming an angle less than 180° at O)

Similarly, computation along OB leads to

$$h[k_1, k_2] = h[2k, -k] = (-1)^k, \qquad k \geq 0$$

Therefore, the values at points on the boundaries of the wedge are

$$h[-k, k] = \left(-\frac{1}{2}\right)^k, \qquad k \geq 0$$

and

$$h[2k, -k] = (-1)^k, \qquad k > 0.$$

The values at the interior of the wedge can subsequently be computed by moving the output mask sequentially along lines parallel to the boundaries. One way of doing

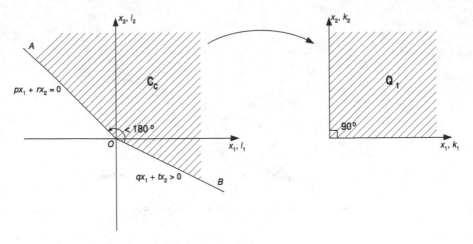

Fig. 4.3 Wedge support to First Quadrant Quarter Plane (FQQP) support

this is as follows. In Fig. 4.2, proceed with the output mask from C downwards to the right along the dashed line indicated. Then, proceed from C' upwards to the left. Subsequently, move sequentially, the mask from D downwards to the right, from D' upwards to the left and continue the procedure.

The transformation from wedge support to first quadrant quarter-plane support is described in Fig. 4.3. Choose p, q, r, t to be non-negative integers such that $pt - qr = 1$. $\mathbb{Z}$ denotes the set of integers, $\mathbb{Z}^2 \triangleq \mathbb{Z} \times \mathbb{Z}$. Then consider the index map

$$\phi : C_c \cap \mathbb{Z}^2 \to Q_1 \cap \mathbb{Z}^2$$

described by

$$\phi[m, n] = [pm + rn, qm + tn]$$

Then,

$$\phi[t, -q] = [pt - qr, 0] = [1, 0]$$
$$\phi[-r, p] = [-pr + pr, -qr + pt] = [0, 1]$$

The fact that $(t, -q)$ and $(-r, p)$ generate $S_{p,q,r,t} = (C_c \cap \mathbb{Z}^2)$ follows from the fact that ϕ given above is a semigroup isomorphism so that for $[m, n] \in S_{p,q,r,t}$, $\phi[m, n] \triangleq [h, k] = h[1, 0] + k[0, 1]$.

The unimodular transformation matrix relates vectors $[k_1, k_2]^T$, $[l_1, l_2]^T$ by

$$\begin{bmatrix} k_1 \\ k_2 \end{bmatrix} = \begin{bmatrix} p & r \\ q & t \end{bmatrix} \begin{bmatrix} l_1 \\ l_2 \end{bmatrix}$$

and it has an inverse,

$$\begin{bmatrix} p & r \\ q & t \end{bmatrix}^{-1} = \begin{bmatrix} t & -r \\ -q & p \end{bmatrix}$$

so that past-present-future points can be defined with respect to a parallelogram, two of whose sides lie on OA and OB. The matrix inverse is associated with the inverse index map

$$\phi^{-1} : Q_1 \cap Z^2 \to C_c \cap Z^2$$

is described by

$$[l_1, l_2]^T = \phi^{-1}[k_1, k_2] = [tk_1 - rk_2, -qk_1 + pk_2].$$

The wedge filter transfer function is of the form

$$H(z_1, z_2) = \sum_S \sum h[\ell_1, \ell_2] z_1^{-\ell_1} z_2^{-\ell_2} \tag{4.9}$$

where S denotes the support of the unit impulse response $\{h[\ell_1, \ell_2]\}$. The FQQP filter, after applying the transformation ϕ is described by

$$\phi H(\alpha, \beta) = \sum_{k_1=0}^{\infty} \sum_{k_2=0}^{\infty} h[tk_1 - rk_2, -qk_1 + pk_2] \, \alpha^{-k_1} \beta^{-k_2}$$

$$= \sum_S \sum h[\ell_1, \ell_2] \, (\alpha^{-1})^{p\ell_1 + r\ell_2} \, (\beta^{-1})^{q\ell_1 + t\ell_2}$$

$$= \sum_S \sum h[\ell_1, \ell_2] \, (\alpha^p \beta^q)^{-\ell_1} \, (\alpha^r \beta^t)^{-\ell_2}. \tag{4.10}$$

Comparing Eqs. (4.9) and (4.10)

$$z_1 = \alpha^p \beta^q, \qquad z_2 = \alpha^r \beta^t \tag{4.11}$$

The preceding set of equations may be inverted and the constraint $pt - qr = 1$ applied to get

$$\alpha = z_1^t z_2^{-q}, \qquad \beta = z_1^{-r} z_2^p \tag{4.12}$$

Thus, if $H(z_1, z_2)$ is a wedge filter transfer function, with the wedge defined by positive integers p, q, r, t, as described above, then $H(z_1^p z_2^q, z_1^r z_2^t)$ is a FQQP filter transfer function.

Consider the wedge filter transfer function expressed in the form

$$H(z_1, z_2) = \frac{1}{1 - C(z_1, z_2)}$$

where $C(z_1, z_2)$ is associated with a recursible output mask. For example, consider the case when

$$H(z_1, z_2) = \frac{1}{0.5 z_1 z_2^{-1} + 1 + 0.85 z_1^{-1} + 0.1 z_1^{-1} z_2^{-1} + 0.5 z_1^{-2} z_2}$$

Then the wedge support of the unit impulse response is determined from a subset of cardinality 2 in the set of indices (k_1, k_2) associated with the monomials $z_1^{-k_1}, z_2^{-k_2}$ in

$$C(z_1, z_2) = -0.5 z_1 z_2^{-1} - 0.85 z_1^{-1} - 0.1 z_1^{-1} z_2^{-1} - 0.5 z_1^{-2} z_2$$

This subset generating the two vectors is shown in Fig. 4.4. The output mask can also be easily generated and has its interior angle at the corner containing the present point to be less than 180°. Clearly, $p = 1, r = 1, q = 1, t = 2, pt - qr = 1$. It is possible to expand $H(z_1, z_2)$ as a formal power series

$$H(z_1, z_2) = \frac{1}{1 - C(z_1, z_2)} = \sum_{k=0}^{\infty} [C(z_1, z_2)]^k$$

It is quite easy to confirm that the support of the wedge filter unit impulse response is contained in the wedge determined by the output mask associated with the denominator of $H(z_1, z_2)$.

$$H(z_1 z_2, z_1 z_2^2) = \frac{1}{1 + 0.5 z_2^{-1} + 0.5 z_1^{-1} + 0.85 z_1^{-1} z_2^{-1} + 0.1 z_1^{-2} z_2^{-3}}$$

4.4 Stability

For stability studies and stability testing (the primary objectives in this chapter) the following fact is necessary.

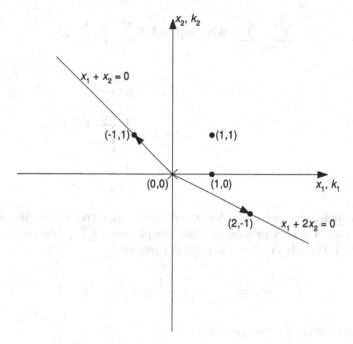

Fig. 4.4 The wedge support for the filter unit impulse response

Fact 4.1. *A 2-D LSI system is BIBO stable*[3] *if and only if its impulse response sequence* $\{h[k_1, k_2]\}$ *is absolutely summable, i.e.,* $\{h[k_1, k_2]\} \in l_1$.

Example 4.3. Consider the z-transform below of the unit impulse response $\{h[k_1, k_2]\}$ of a filter

$$H(z_1, z_2) = 1 + \sum_{k=1}^{\infty} \frac{1}{100k^3}(z_2^k + z_2^{-k})z_1^{-1}$$

$$h[k_1, k_2] = \begin{cases} 1, & k_1 = k_2 = 0 \\ \frac{1}{100|k_2|^3}, & k_1 = 1, k_2 \neq 0 \\ 0, & \text{otherwise} \end{cases}$$

To determine if the filter is BIBO stable, proceed as follows.

---

[3]Editorial note: The concept of Bounded-Input-Bounded-Output (BIBO) stability is more problematic and its use far more controversial in dimensions higher than one (cf. brief note at the end of this chapter). It may be noted that BIBO stability has not been defined in this text, but FACT 4.1 can itself be adopted as a definition. Alternatively, Definition 3.7 from the 1982 version of this book may be adopted, which essentially states that a system is BIBO stable if any bounded input $\{x[k_1, k_2]\} \in \ell_\infty$ produces a bounded output $\{y[k_1, k_2]\} \in \ell_\infty$.

$$\sum_{k_1=-\infty}^{\infty} \sum_{k_2=-\infty}^{\infty} |h[k_1,k_2]| = 1 + \sum_{k_2 \neq 0} \frac{1}{100|k_2|^3}$$

$$= 1 + \frac{1}{50} \sum_{k=1}^{\infty} \frac{1}{k^3}$$

$$< 1 + \frac{1}{50} \sum_{k=1}^{\infty} \frac{1}{k^2}$$

$$< 1 + \frac{1}{50} \frac{\pi^2}{6} < \infty$$

The unit impulse response is absolutely summable and, therefore, the filter is BIBO stable. In fact, it is easy to show that the infinite series $\sum_{k=1}^{\infty} \frac{1}{k^p}$ converges if and only if $p > 1$. For this, consider the improper integral,

$$\int_1^\infty \frac{1}{x^p} dx = \lim_{b \to \infty} \frac{b^{1-p} - 1}{1-p} = \frac{1}{p-1}, \quad p > 1$$

The improper integral converges when $p > 1$.

Rigorous proof of Fact 4.1 can be given for the first quadrant quarter-plane case. In [62] the stability criterion of n-dimensional filters is derived in a somewhat general setting. The numerator and the denominator of the filter transfer function are each considered to be power series with absolutely summable coefficients, i.e., the sequence of coefficients in each power series is assumed to be in l_1. Then, for an input sequence in l_1, conditions are investigated for the output sequence to be in l_1 in terms of absolutely summable denominator power series of the filter. It is noted that the convolution of two sequences in l_1 is also in l_1. (The convolution of two sequences is associated with the operation of multiplication of their z-transforms.) Consider index sets belonging to the set $S = Z^\alpha * P^\beta * N^\gamma$, where α, β, γ are nonnegative integers and Z, P, N, respectively, represent the sets of integers, nonnegative integers, and non-positive integers. The denominator power series

$$B(z_1, \cdots, z_n) = \sum_{(k_1, \cdots, k_n) \in S} \cdots \sum b[k_1, k_2, \cdots, k_n] z_1^{k_1} z_2^{k_2} \cdots z_n^{k_n}$$

is the z-transform of the sequence $\{b[k_1, \cdots, k_n]\}$. Let $l_1(S)$ denote the sequences $\{b[k_1, \cdots, k_n]\}$ defined on S which satisfy, $\sum_{k_1} \cdots \sum_{k_n} |b[k_1, \cdots, k_n]| < \infty$, with convolution as multiplication. If $\{b[k_1, \cdots, k_n]\} \in l_1(S)$, then $1/B(z_1, \cdots, z_n)$ is stable if and only if there exists $\{a[k_1, \cdots, k_n]\} \in l_1(S)$ such that $A(z_1, \cdots, z_n)B(z_1, \cdots, z_n) = 1$, where $A(z_1, \cdots, z_n)$ is the z-transform of $\{a[k_1, \cdots, k_n]\}$; that is, the problem of stability is equated to the problem of invertibility in $l_1(S)$. The space $l_1(S)$ is a commutative Banach algebra with identity (with convolution as multiplication) and its dual is known to be $l_\infty(S)$. The main result will be given as Theorem 4.1 here; it rests on the use of a Tauberian theorem proved by Wiener, which is stated first.

Fact 4.2. *Let C be a commutative Banach algebra with identity, and let C_1 be its dual. An element $c \in C$ is invertible if and only if the equation $\phi(c) = 0$ is not satisfied by any homomorphism ϕ in C_1.*

Theorem 4.1. *Let $S = Z^\alpha * P^\beta * N^\gamma, \alpha + \beta + \gamma > 0$ and let $\{b[k_1, \cdots, k_n]\} \in l_1(S)$, where $k_1, \cdots, k_n \in S$. Then the filter with transfer function $1/B(z_1, \cdots, z_n)$ (where $B(z_1, \cdots, z_n)$ is the z-transform of $\{b[k_1, \cdots, k_n]\}$ has a convolution inverse in $l_1(S)$ if and only if $B(z_1, \cdots, z_n) \neq 0$ for*

$$|z_j| \begin{cases} = 1 \text{ , if } R_j = Z \\ \leq 1 \text{ , if } R_j = P \\ \geq 1 \text{ , if } R_j = N \end{cases}$$

and $1 \leq j \leq n$, where R_j is the range of k_j.

The proof of Theorem 4.1 can be found elsewhere. A special case, quite commonly encountered, will be stated and proved later in Theorem 4.2, using elementary arguments. However, several consequences of the result just given are worth discussing. The following definition is necessary.

Definition 4.4. An n-D LSI system will be called l_p-stable, $1 \leq p \leq \infty$, provided any input sequence $\{x[k_1, \cdots, k_n]\} \in l_p$ to the system produces an output sequence $\{y[k_1, \cdots, k_n]\} \in l_p$.

BIBO stability is, thus, equivalent to l_∞-stability in the sense of this definition. Since the convolution of two sequences, each in $l_1(S)$, is also in $l_1(S)$ and the convolution of a sequence in $l_\infty(S)$ with a sequence in $l_1(S)$ is in $l_\infty(S)$, it follows from Fact 4.1 that for LSI systems the conditions for l_1-stability and l_∞-stability are identical (only a finite input sequence to the system need be considered). In fact it has been concluded that for time-invariant systems l_1-stability, l_∞-stability, and l_p-stability, $1 < p < \infty$, conditions are all equivalent and are each equivalent to the condition that the impulse response sequence of the system belong to l_1. In addition, it has been shown in [63] that l_1-stability and l_∞-stability together imply l_p-stability, $1 \leq p \leq \infty$, for very general classes of systems. The usefulness of Theorem 4.1 arises from the realization that a general stability criterion based on the requirement that the impulse response sequence be in $l_1(S)$ becomes verifiable via algebraic tests for a broad class of filters encountered in practice. Note that when a filter characterized by $1/B(z_1, \cdots, z_n)$ satisfies the appropriate stability condition, depending on the index set S, the change from an unity numerator to any other with coefficient sequence in $l_1(S)$ does not alter the stability property. However, as will be substantiated in the next section, stable filters exist with a nonunity numerator and a denominator not satisfying the appropriate condition in Theorem 4.1.

Finally, some caution must be exercised in interpretation and application of Theorem 4.1. The support of $\{b[k_1, \cdots, k_n]\}$ may be different from the support of

$\{a[k_1, \cdots, k_n]\}$ whose z-transform $A(z_1, \cdots, z_n)$ is the inverse of the z-transform of $\{b[k_1, \cdots, k_n]\}$. This is the case when one or more of the indices of $\{b[k_1, \cdots, k_n]\}$ have a range $R_j \subset Z$.

4.5 Stability Properties of Quarter-Plane Filters

The impulse response of a quarter-plane filter has support in one of the four quadrants. For first quadrant quarter-plane filters, (4.5) specializes to

$$y[k_1, k_2] = \sum_{i_1=0}^{m_1} \sum_{i_2=0}^{m_2} a[i_1, i_2] x[k_1 - i_1, k_2 - i_2]$$

$$- \sum_{\substack{i_1 = 0 \\ i_1 + i_2 \neq 0}}^{n_1} \sum_{i_2=0}^{n_2} b[i_1, i_2] y[k_1 - i_1, k_2 - i_2] \qquad (4.13)$$

The recursion equations for the second, third, and fourth quadrant filters are similar to (4.13), except that $y[k_1 - n_1, k_2]$, $y[k_1 - n_1, k_2 - n_2]$, $y[k_1, k_2 - n_2]$, respectively, appear on the left-hand side of the equation and $b[n_1, 0]$, $b[n_1, n_2]$, $b[0, n_2]$ are all nonzero (and can set equal to 1, like $b[0, 0] = 1$ in (4.13). The impulse response of the ith quadrant quarter-plane filter is in the ith quadrant for $i = 1, 2, 3, 4$. The two most common orders of computation for the first quadrant filter are row by row from left to right and bottom to top, and column by column from bottom to top and left to right. Of course, the output sequence is independent of the chosen order of computation. First quadrant filters are said to recurse in the $++$ direction and, second, third, and fourth quadrant filters are said to recurse, respectively, in the $-+$, $--$, and $+-$ directions.

On taking the 2-D z-transform of (4.13) with $b[0, 0] = 1$, and zero boundary conditions, one obtains

$$H(z_1, z_2) = \frac{Y(z_1, z_2)}{X(z_1, z_2)} = \frac{\sum_{i_1=0}^{m_1} \sum_{i_2=0}^{m_2} a[i_1, i_2] z_1^{i_1} z_2^{i_2}}{1 + \sum_{i_1=0, i_1+i_2 \neq 0}^{n_1} \sum_{i_2=0}^{n_2} b[i_1, i_2] z_1^{i_1} z_2^{i_2}} \qquad (4.14)$$

Since the denominator polynomial of $H(z_1, z_2)$ in (4.14) is nonzero in some neighborhood $\{(z_1, z_2) : |z_1| < \epsilon |z_2| < \epsilon\}$ of $(0, 0)$, the function $H(z_1, z_2)$ is analytic and has a power series expansion in such a neighborhood:

$$H(z_1, z_2) = \sum_{i_1=0}^{\infty} \sum_{i_2=0}^{\infty} h[i_1, i_2] z_1^{i_1} z_2^{i_2} \qquad (4.15)$$

It is necessary only to study the stability properties of first quadrant filters because of the following result.

Fact 4.3. *A transfer function $H(z_1, z_2)$ can be realized as a stable second, third, or fourth quadrant filter if and only if $H(z_1^{-1}, z_2), H(z_1^{-1}, z_2^{-1})$, or $H(z_1, z_2^{-1})$, respectively, can be realized as a stable first quadrant filter.*

4.5.1 Transform Domain Formulation

Proper subset of the unit polydisc, in the case $n > 1$, is defined as

$$\Gamma \triangleq \{(z_1, \cdots, z_n) : |z_1| \le 1, |z_2| = 1, \cdots, |z_n| = 1\}$$

The following stability theorem is important.

Theorem 4.2. *The 2-D system described by the transfer function in (4.14) is BIBO stable if*

$$B(z_1, z_2) \triangleq 1 + \sum_{\substack{i_1 = 0 \\ i_1 + i_2 \ne 0}}^{n_1} \sum_{i_2=0}^{n_2} b[i_1, i_2] z_1^{i_1} z_2^{i_2} \ne 0, \quad in \ \ \overline{U}^2 \tag{4.16}$$

Proof. If $B(z_1, z_2) \ne 0$ in $\overline{U}^2$, then by continuity of $B(z_1, z_2)$ there exists a larger open bidisc $U_{1+\epsilon}^2 \triangleq \{(z_1, z_2) : |z_1| < 1 + \epsilon, |z_2| < 1 + \epsilon, \epsilon > 0\}$ such that $B(z_1, z_2) \ne 0$ in $U_{1+\epsilon}^2$. Consequently, $H(z_1, z_2)$ in (4.14) is analytic in $U_{1+\epsilon}^2$, implying that $\sum_{i_1=0}^{\infty} \sum_{i_2=0}^{\infty} h[i_1, i_2] z_1^{i_1} z_2^{i_2}$ in (4.15) is absolutely convergent in $U_{1+\epsilon}^2$. This implies that the sequence $\{h[k_1, k_2]\} \in l_1$, and the proof is complete.

Theorem 4.2 was given earlier by Shanks, except for the fact that the condition $B(z_1, z_2) \ne 0$ in $\overline{U}^2$ for unity numerator was also necessary. It has been shown subsequently that with non-unity numerators,

$$H_1(z_1, z_2) \triangleq \frac{A_1(z_1, z_2)}{B(z_1, z_2)} = \frac{(1 - z_1)^8 (1 - z_2)^8}{2 - z_1 - z_2} \tag{4.17}$$

is BIBO stable, while

$$H_2(z_1, z_2) \triangleq \frac{A_2(z_1, z_2)}{B(z_1, z_2)} = \frac{(1 - z_1)(1 - z_2)}{2 - z_1 - z_2} \tag{4.18}$$

is BIBO unstable, even though in each case the denominator polynomial $B(z_1, z_2) \ne 0$ in $\overline{U}^2$ except at $z_1 = z_2 = 1$, at which point both $H_1(z_1, z_2)$ and $H_2(z_1, z_2)$

have nonessential singularities of the second kind. It is, therefore, apparent that, in contrast to the 1-D case, the numerator polynomial of the filter transfer function can influence the BIBO stability property in the $n > 1$ case. A necessary condition for BIBO stability is stated next.

Fact 4.4. *If $H(z_1, z_2)$ is the transfer function of a BIBO stable 2-D LSI filter, then $H(z_1, z_2)$ has no poles in $\overline{U}^2$ (i.e., no nonessential singularities of the first kind (NSFK) in $\overline{U}^2$) and no nonessential singularities of the second kind (NSSK) in $\overline{U}^2$ except possibly on T^2.*

Recall that the number of NSSK's of a bivariate rational function with coprime numerator and denominator is always finite. It is possible to establish whether or not nonessential singularities of the second kind on T^n exist in an n-D filter transfer function $H(z_1, \cdots, z_n), n > 1$. However, resolution of the stability problem in general due to the presence of such singularities remains difficult. The test for $B(z_1, z_2) \neq 0$ in $\overline{U}^2$ in (4.16) is simplified via the result given next.

Theorem 4.3. *The bivariate polynomial $B(z_1, z_2) \neq 0$ in $\overline{U}^2$ if and only if:*

$$(i)\, B(0, z_2) \neq 0, \quad |z_2| \leq 1$$

$$(ii)\, B(z_1, z_2) \neq 0, \quad |z_1| \leq 1, |z_2| = 1$$

Proof. The "only if" part obviously holds, and the "if" part is proved here. Let $z_1 = f(z_2)$ be the algebraic function obtained from $B(z_1, z_2) = 0$. Condition (i) implies that $f(z_2) \neq 0$ for $|z_2| \leq 1$. It is known that for a nonconstant algebraic function $f(z_2) \neq 0$ for $|z_2| \leq 1$, the minimum value of the modulus $|f(z_2)|$ for $|z_2| \leq 1$ cannot occur when $|z_2| < 1$. This coupled with the implication of condition (ii) that $|f(z_2)| > 1$ for $|z_2| = 1$ leads to the fact the $|f(z_2)| > 1$ if $|z_2| \leq 1$. Hence $B(z_1, z_2) \neq 0$ in $\overline{U}^2$.

Alternative Proof. Again, the proof for the "if" part is considered, as the "only if" part is trivial. Define

$$N(z_1) = \frac{1}{2\pi j} \oint_{|z_2|=1} \frac{\partial B(z_1, z_2)}{\partial z_2} [B(z_1, z_2)]^{-1} dz_2$$

For any fixed $z_1 = z_{10}$ in $|z_1| \leq 1, N(z_{10})$ is the number of z_2-zeros of $B(z_{10}, z_2)$ in $|z_2| \leq 1$. Condition (ii) implies that $N(z_1)$ is continuous in $|z_1| \leq 1$. Also, for any fixed $z_1, N(z_1)$ is integer-valued, and therefore must be a constant in $|z_1| \leq 1$. Condition (i) implies that $N(0) = 0$. Hence $N(z_1) = 0, |z_1| \leq 1$, implying that $B(z_1, z_2) \neq 0$ in $\overline{U}^2$, as was to be proved.

The first proof for the Theorem 4.3 is based on arguments advanced by Davis. Other proofs have also been advanced. Several comments are in order. First, the theorem holds for functions analytic on U^2 and continuous on $\overline{U}^2$, and not merely for polynomials. Second, the theorem is the key to the proof by induction of a similar result valid for n-variate polynomials, stated next.

4.6 Algebraic 2-D Stability Tests

The test requires the polynomial $B(w_1, w_2)$ in the complex variables w_1 and w_2 to be tested for absence of zeros in $\overline{U}^2 : \{(w_1, w_2) : |w_1| \leqslant 1 \text{ and } |w_2| \leqslant 1\}$. It has been established that

$$B(w_1, w_2) \neq 0 \text{ in } \overline{U}^2 \longleftrightarrow \text{ (a) } B(w_1, 0) \neq 0, |w_1| \leqslant 1 \text{ and}$$
$$\text{(b) } B(w_1, w_2) \neq 0, |w_1| = 1, |w_2| \leqslant 1$$

Test for criterion (a) in the preceding equivalence requires the application of a 1-D test procedure. As a digression, one such procedure is reviewed for generalization to test for criterion (b). Now,

$$\sum_{i=0}^{m} a_i w_1^i \neq 0, |w_1| \leqslant 1 \longleftrightarrow \sum_{i=0}^{m} a_{m-i} w_1^i \neq 0, |w_1| \geqslant 1$$

where, the coefficients, in general are complex-valued.

The generating rows are the first two rows in the table below and the first column is the pivot column.

	Pivot Column						N & S Condition
	a_0	a_1	a_2	$\cdots$ a_{m-2} a_{m-1} a_m			$\left|\frac{a_m}{a_0}\right| < 1$
	a_m^*	a_{m-1}^*	a_{m-2}^*	$\cdots$ a_2^* a_1^* a_0^*			
Generated	b_0	b_1	b_2	$\cdots$ b_{m-2} b_{m-1}			$\left|\frac{b_{m-1}}{b_0}\right| < 1$
Rows	b_{m-1}^*	b_{m-2}^*	b_{m-3}^*	$\cdots$ b_1^* b_0^*			
	c_0	c_1	c_2	$\cdots$ c_{m-2}			$\left|\frac{c_{m-2}}{c_0}\right| < 1$
	c_{m-2}^*	c_{m-3}^*	c_{m-4}^*	$\cdots$ c_0^*			
	$\vdots$	$\vdots$	$\vdots$				$\vdots$

Each generated row has elements from left to right obtained by taking 2×2 matrix determinants of matrices formed with the pivot column and one other column, starting from the last. Thus,

$$b_0 = \begin{vmatrix} a_0 & a_m \\ a_m^* & a_0^* \end{vmatrix} = |a_0|^2 - |a_m|^2, b_1 = \begin{vmatrix} a_0 & a_{m-1} \\ a_m^* & a_1^* \end{vmatrix}, \cdots, b_{m-1} = \begin{vmatrix} a_0 & a_1 \\ a_m^* & a_{m-1}^* \end{vmatrix}$$

and so on. To test for $B(w_1, w_2) \neq 0, |w_1| \leq 1, |w_2| \leq 1$, first write $B(w_1, w_2)$ in recursive canonical form in the main variable w_2 i.e.

$$B(w_1, w_2) = \sum_{k=0}^{n} a_k(w_1)w_2^k$$

Form the generating rows as in the 1-D case.

$$\frac{a_0(w_1) \quad a_1(w_1) \quad a_2(w_1) \quad \cdots \quad a_{n-2}(w_1) \quad a_{n-1}(w_1) \quad a_n(w_1)}{a_n(w_1) \quad a_{n-1}(w_1) \quad a_{n-2}(w_1) \quad \cdots \quad a_2(w_1) \quad \quad a_1(w_1) \quad a_0(w_1)}$$

The overbar denotes complex conjugation of the term below it. Note that on $|w_1| = 1, \overline{w_1} = w_1^{-1}$. Therefore, the first element of the first generated row is

$$b_0(w_1, \overline{w_1}) = a_0(w_1)\overline{a_0(w_1)} - a_n(w_1)\overline{a_n(w_1)}$$

which is always real-valued. Therefore, on $|w_1| = 1, b_0(w_1, w_1^{-1})$ must be expressible in the form

$$b_0(w_1, \overline{w_1}) = b_0(w_1, w_1^{-1}) = \sum_{k=0} b_{1k}(w_1^k + w_1^{-k})$$

$$= \sum_{k=0} b_{1k} \cos k\theta$$

where w_1 may be parameterized on $|w_1| = 1$ as $e^{j\theta}$ so that $(w_1 + w_1^{-1})/2 = \cos\theta \triangleq x_1$.

The Chebyshev polynomial of the first kind and n^{th} order is

$$T_n(x) = \cos(n \cos^{-1} x) = \cos n\theta, \theta = \cos^{-1} x$$

The recurrence relation below generates all $T_n(x), n > 1$.

$$T_{n+1}(x) - 2xT_n(x) + T_{n-1}(x) = 0, T_0(x) = 1, T_1(x) = x$$

Therefore,

$$b_0(w_1, w_1^{-1}) = \sum_{k=0} b_{1k}T_n(x)$$

Since $\left|\frac{a_n(w_1)}{a_0(w_1)}\right| < 1$ on $|w_1| = 1 \longleftrightarrow -1 \leqslant x \leqslant 1$, all we need to do is test for positivity: $\sum_{k=0} b_{1k}T_n(x) > 0, -1 \leqslant x \leqslant 1$. This is easy to do and is repeated on the first element of each generated row.

Example 4.4.

$$A(w_1, w_2) = 12 + 6w_1 + 10w_2 + 5w_1w_2 + 2w_2^2 + w_1w_2^2$$

Is $A(w_1, w_2) \neq 0, |w_1| \leqslant 1, |w_2| \leqslant 1$

$$A(w_1, 0) = 12 + 6w_1 \neq 0 |w_1| \leqslant 1 \qquad\qquad\qquad \text{EASY}$$

Condition (ii) is now to be checked

$$A(w_1, w_2) = (12 + 6w_1) + (10 + 5w_1)w_2 + (2 + w_1)w_2^2$$

$$= \sum_{k=0}^{2} a_k(w_1)w_2^k$$

where $a_0(w_1) = 12 + 6w_1$, $a_1(w_1) = 10 + 5w_1$, $a_2(w_1) = 2 + w_1$. Recalling that on $|w_1| = 1, w_1^* = w_1^{-1}$, the generating rows are

$$\begin{array}{ccc} 12 + 6w_1 & 10 + 5w_1 & 2 + w_1 \\ 2 + w_1^{-1} & 10 + 5w_1^{-1} & 12 + 6w_1^{-1} \end{array}$$

$$\begin{array}{cc} b_0(w_1, w_1^{-1}) & b_1(w_1, w_1^{-1}) \\ b_1(w_1), w_1^{-1} & b_0(w_0, w_1^{-1}) \end{array}$$

$$B_0\overline{B_0} - B_1\overline{B_1}$$

where $B_0 = b_0(w_1, w_1^{-1}) \triangleq (12 + 6w_1)(12 + 6w_1^{-1}) - (2 + w_1)(2 + w_1^{-1})$ and $B_1 = b_1(w_1, w_1^{-1}) \triangleq (12 + 6w_1)(10 + 5w_1^{-1}) - (2 + w_1^{-1})(10 + 5w_1)$. Now,

$$b_0(w_1, w_1^{-1}) = 144 - 4 + 36 - 1 + 72(w_1 + w_1^{-1}) - 2(w_1 + w_1^{-1})$$

$$= 175 + 140x_1, \text{ where } x_1 = T_1(x_1) = \frac{w_1 + w_1^{-1}}{2}$$

Clearly, $175 + 140x_1 > 0, -1 \leqslant x_1 \leqslant 1$. Similarly, noting that $B_1 = 125 + 100x_1$,

$$B_0\overline{B_0} - B_1\overline{B_1} = (175 + 140x_1)(175 + 140x_1) - (125 + 100x_1)(125 + 100x_1)$$

$$\therefore B_0\overline{B_0} - B_1\overline{B_1} > 0, -1 \leqslant x_1 \leqslant 1$$

$$\therefore A(w_1, w_2) \neq 0, |w_1| = 1, |w_2| \leqslant 1$$

Finally, $A(w_1, w_2) \neq 0, |w_1| \leqslant 1, |w_2| \leqslant 1$.

Fact 4.5. *The n-variate polynomial $B(z_1, \cdots, z_n) \neq 0$ in $\overline{U}^2$ if and only if:*

(a) $B(0, 0, \cdots, 0, z_n) \neq 0, |z_n| \leq 1$
(b) $B(0, 0, \cdots, z_{n-1}, z_n) \neq 0, |z_{n-1}| \leq 1, |z_n| = 1$

$$\vdots$$

(m) $B(0, z_1, \cdots, z_{n-1}, z_n) \neq 0, |z_2| \leq 1, \bigcap_{k=3}^{n}\{|z_k| = 1\}$
(n) $B(z_1, z_2, \cdots, z_{n-1}, z_n) \neq 0, |z_1| \leq 1, \bigcap_{k=2}^{n}\{|z_k| = 1\}$

A more general version of Theorem 4.3 is stated and proved next.

Theorem 4.4. *The bivariate polynomial $B(z_1, z_2) \neq 0$ in $\overline{U}^2$ if and only if:*

(i) $B(z_{10}, z_2) \neq 0$, *for any* $z_1 = z_{10}, |z_{10}| \leq 1, |z_2| \leq 1$
(ii) $B(z_1, z_{20}) \neq 0$, *for* $|z_1| \leq 1$, *for any* $z_2 = z_{20}, |z_{20}| = 1$
(iii) $B(z_1, z_2) \neq 0$, *in* $T^2 \triangleq \{(z_1, z_2) : |z_1| = 1, |z_2| = 1\}$.

Proof. The proof for the "only if" part is trivial. To prove the "if" part, let $N(z_{20})$ be the number of zeros in z_1 of $B(z_1, z_{20})$ that fall in $|z_1| \leq 1$. Then, from (ii) and (iii), one has

$$N(z_{20}) = \frac{1}{2\pi j} \oint_{|z_1|=1} \frac{\partial B(z_1, z_{20})}{\partial z_1} [B(z_1, z_{20})]^{-1} dz_1 \qquad (4.19)$$

For arbitrary but fixed z_2 on $|z_2| = 1$, an expression for $N(z_2)$ can be written similar in form to (4.19). It is possible to conclude (as in Theorem 4.3) that $N(z_2)$ is continuous on $|z_2| = 1$, which along with the fact that $N(z_2)$ is integer-valued for any fixed z_2 implies that $N(z_2) = N(z_{20}) = 0$ (from (ii), $N(z_{20}) = 0$), $|z_2| = 1$. Therefore

$$b(z_1, z_2) \neq 0, \quad |z_1| \leq 1, |z_2| = 1.$$

This is identical to condition (ii) in Theorem 4.3. Also, condition (i) is identical to condition (i) in Theorem 4.3. The proof of the theorem is complete.

Example 4.5.

$$N(w_1, w_2) = w_2^3 + 2w_2w_1 + (1 + w_2)w_1^2 \in R[w_2][w_1]$$

$$D(w_1, w_2) = -3w_2^2 - 6w_1 + w_1^2 \in R[w_2][w_1]$$

$$\underline{R}(w_2) = \begin{bmatrix} 1 + w_2 & 2w_2 & w_2^3 & 0 \\ 0 & 1 + w_2 & 2w_2 & w_2^3 \\ 0 & 1 & -6 & -3w_2^2 \\ 1 & -6 & -3w_2^2 & 0 \end{bmatrix}$$

$\det \underline{R}(w_2) = -w_2^4(4w_2 + 3)^2 = 0 \Rightarrow w_2 = 0, -\frac{3}{4}$.

For $N = 0, D = 0$, we must have $w_2 = 0$, or $w_2 = -\frac{3}{4}$.

When $w_2 = 0$, $w_1^2 - 6w_1 = 0$ and $w_1^2 = 0$, as $w_1 = 0$.

When $w_2 = -\frac{3}{4}$,

$$\frac{w_1^2}{4} - \frac{3w_1}{2} - \frac{27}{64} = 0$$

$$w_1^2 - 6w_1 - \frac{27}{16} = 0$$

$$\rightarrow w_1^2 - 6w_1 - \frac{27}{16} = 0 \quad \rightarrow w_1 = 3 \pm \frac{3\sqrt{19}}{4}$$

Therefore NSSK are at $(0, 0)$, $(3 + \frac{3}{4}\sqrt{19}, -\frac{3}{4})$ and $(3 - \frac{3}{4}\sqrt{19}, -\frac{3}{4})$.

Example 4.6.

$$D(z_1^{-1}, z_2^{-1}) = 1 - 0.5z_1^{-1} - 1.5z_2^{-1} + 1.8z_1^{-1}z_2^{-1} + 0.6z_2^{-2} - 0.72z_1^{-1}z_2^{-2}$$
$$+ 0.5z_1^{-2} - 0.75z_1^{-2}z_2^{-1} + 0.29z_1^{-2}z_2^{-2}$$
$$D(z_1^{-1}, 0) = 1 - 0.5z_1^{-1} + 0.5z_1^{-2} \neq 0, \quad |z_1^{-1}| \leq 1$$
$$C(z_2) = z_2^2 D(z_1^{-1}, z_2^{-1}) = (0.6 - 0.72z_1^{-1} + 0.29z_1^{-2}) +$$
$$(-1.5 + 1.8z_1^{-1} - 0.75z_1^{-2})z_2 +$$
$$(1 - 1.2z_1^{-1} + 0.5z_1^{-2})z_2^2$$

Schur-Chon-Matrix on $|z_1| = 1$

$$C(z_1) = \begin{bmatrix} c_{11}(z_1) & c_{12}(z_1) \\ \hline c_{21}(z_1) = c_{12}^*((z_1^*)^{-1}) = [c_{12}((z_1^*)^{-1})]^* & c_{11}(z_1) \end{bmatrix}$$

$$C(z_1) = \begin{bmatrix} \begin{matrix} 1.7275 - 1.1592(z_1 + z_1^{-1}) \\ +0.326(z_1^2 + z_1^{-2}) \end{matrix} & \begin{matrix} -0.3z_1^{-2} + 1.08z_1^{-1} - 1.6215 \\ +1.098z_1 - 0.315z_1^2 \end{matrix} \\ \hline \begin{matrix} -0.3z_1^2 + 1.08z_1 - 1.6215 \\ +1.098z_1^{-1} - 0.315z_1^{-2} \end{matrix} & \begin{matrix} 1.7275 - 1.1592(z_1 + z_1^{-1}) \\ +0.326(z_1^2 + z_1^{-2}) \end{matrix} \end{bmatrix}$$

$$C(1) = \begin{bmatrix} 0.0611 & -0.0585 \\ -0.0585 & 0.0611 \end{bmatrix} > 0$$

$$\det C(z_2) = 0.6938 - 0.5593(z_1 + z_1^{-1}) + 0.287(z_1^2 + z_1^{-2}) -$$
$$0.0862(z_1^3 + z_1^{-3}) + 0.0118(z_1^4 + z_1^{-4})$$

$$x_1 = \frac{z_1 + z_1^{-1}}{2} = \cos\theta$$

$$\det C(z_2) = 0.6938 - 2[0.5593(T_1(x_1)) + 0.287(T_2(x_1)) -$$
$$0.0862(T_3(x_1)) + 0.0118(T_4(x_1))]$$
$$= 0.143334 - 0.60148x_1 + 0.95963x_1^2 - 0.689587x_1^3 + 0.188416x_1^4$$

(since $T_1(x) = x$, $T_2(x) = 2x^2 - 1$, $T_3(x) = 4x^3 - 3x$, $T_4(x) = 8x^4 - 8x^2 + 1$)
Roots of $C(z_1)$ on $|z_1| = 1$ occur for values of x_1 at:

$$0.912134 \pm j0.214885, \quad 0.917824 \pm j0.154509$$

(Using Mathematica; you may use MAPLE(VAX), MATLAB, etc.)
Therefore $\underline{C}(z_1)$ is positive definite on $|z_1| = 1$. FILTER STABLE
(b) Same as (a) except $0.29z_1^{-2}z_2^{-2} \to 0.25z_1^{-2}z_2^{-2}$
By Mathematica, the elements of the Schur-Cohn matrix are

$$C_{11}(z_1) = 1.7491 - 1.188(z_1 + z_1^1) + 0.35(z_1^2 + z_1^{-2}), \quad |z_1| = 1$$
$$C_{12}(z_1) = -0.3z_1^{-2} + 1.08z_1^{-1} - 1.6515 + 1.17z_1 - 0.375z_1^2 = C_{21}(z_1^{-1})$$
$$C_{22}(z_1) = C_{11}(z_1)$$
$$\det \underline{C}(z_1) = 0.633662 - 0.508837(z_1 + z_1^{-1}) + 0.257351(z_1^2 + z_1^{-2}) -$$
$$0.0756(z_1^3 + z_1^{-3}) + 0.01(z_1^4 + z_1^{-4})$$
$$= 0.13896 - 0.564074x + 0.869404x^2 - 0.6048x^3 + 0.16x^4$$

Roots on $|z_1|$ occur at values of $x = \frac{z_1 + z_1^{-1}}{2} = \cos\theta$ at $0.799659 \in [-1, 1]$, $0.919511 \pm j0.325741, 1.14132$. Therefore $\underline{C}(z_1)$ is not positive definite on $|z_1| = 1$. UNSTABLE FILTER

The proof of Theorem 4.4 can also be given via use of algebraic function theory instead of the Cauchy principal value theorem. A proof based on algebraic topological approach using arguments from homotopy theory[4] has also been given. The generalization of Theorem 4.4 to n dimensions is stated next.

Fact 4.6. *The n-variate polynomial* $B(z_1, \cdots, z_n) \neq 0$ *in* $\overline{U}^2$ *if and only if:*

(i) *For any* $z_{10}, z_{20}, \cdots, z_{n0}$ *such that* $|z_{r0}| = 1, r = 1, \cdots, n$, *and for all* $i, i = 1, \cdots, n, B(z_1, \cdots, z_n) \neq 0$, *when* $z_{r0} = z_r, r \neq i, |z_i| \leq 1$.

(ii) $B(z_1, \cdots, z_n) \neq 0$, *in* $T^n \triangleq \{(z_1, \cdots, z_n) : |z_1| = 1, \cdots, |z_n| = 1\}$.

A variant of Fact 4.6 is given next, where a test for zeros in the distinguished boundary of the polydisc is required but the n single-variable tests are replaced by one single-variable test. (Of course, the polynomial, in general will be of higher degree.) The proof uses arguments from homotopy theory and is given elsewhere.

Fact 4.7. *The n-variate polynomial* $B(z_1, \cdots, z_n) \neq 0$ *in* $\overline{U}^2$ *if and only if:*

$$(i) \quad B(z_1, z_1, \cdots, z_1) \neq 0, \qquad |z_1| \leq 1$$
$$(ii) \quad B(z_1, z_2, \cdots, z_n) \neq 0, \qquad on \quad T^n.$$

In fact, condition (i) of Fact 4.7 can be replaced by

$$B(z_1^{k_1}, z_1^{k_2}, \cdots, z_1^{k_n}) \neq 0, \quad |z_1| \leq 1$$

for any fixed integer $k_i > 0, i = 1, \cdots, n$. It may be noted that Fact 4.7 follows from Fact 4.6, using a specialization of a result, which states that when $B(z_1, z_2, \cdots, z_n) \neq 0$ on T^n and $B_j(z_1) = B(1, 1, \cdots, 1, z_1, 1, \cdots, 1)$ (z_1 in the jth place) and k_j is the number of zeros of $B_j(z_1)$ in $|z_1| \leq 1$, then $B(z_1, z_1, \cdots, z_1)$ has $k_1 + k_2 + \cdots + k_n$ zeros in $|z_1| \leq 1$.

Before discussing another procedure to test for absence of zeros of $B(z_1, z_2)$ in $\overline{U}^2$, some facts concerning 1-D sequences will be introduced. Assume that $b(k_1)$ is a finite sequence of real numbers, $b(\alpha), \cdots, b(\beta)$, with $\alpha \leq \beta, \beta > 0$, and $b(\alpha) \neq 0, b(\beta) \neq 0$. A polynomial possibly in z_1, z_1^{-1} (z_1^{-1} powers are present when $\alpha < 0$), $B(z_1, z_1^{-1}) \triangleq \sum_{k_1=\alpha}^{\beta} b(k_1) z_1^{k_1}$ can be formed which has $m_1 = m_{1i} + m_{10}$ zeros (counting multiple zeros and zeros at infinity), where

$$m_{1i} = |\min(0, \alpha)|, \qquad m_{10} = \max(0, \beta).$$

---

[4]Editorial note: Use of homotopy in such contexts is a manifestation of the simple fact that zeros of a polynomial are continuous functions of coefficients of the polynomial. More comments on notes at the end of chapter.

When $B_1(z_1, z_1^{-1})$ has no zeros on $|z_1| = 1$, it is possible to define on $[-\pi, pi]$ a continuous odd phase function associated with the Fourier transform of the sequence. This phase function contains no linear phase component if and only if $B_1(z_1, z_1^{-1})$ has exactly m_{1i} zeros inside and m_{10} outside $|z_1| = 1$. If the unit circle $|z_1| = 1$ is defined to be a Nyquist contour, the preceding phase condition implies that the Nyquist plot does not encircle or pass through the origin. The following result is true.

Fact 4.8. *The bivariate polynomial $B(z_1, z_2) \neq 0$ in $\overline{U}^2$ if and only if: (i) $B(z_1, 1)$ and $B(1, z_2)$ have no linear phase terms on $|z_1| = 1, |z_2| = 1$, respectively. (ii) $B(z_1, z_2) \neq 0$ on T^2.*

This result is true for any type of recursively computable filter, with output masks of various shapes. More will be said about the result in the discussion of the stability of a general recursive filter. It has been shown that these conditions are equivalent to the imposition of certain restrictions on the phase of the Fourier transform of the array $\{b[k_1, k_2]\}$ with which $B(z_1, z_2)$ is associated. Let

$$B_{\omega_1} = \sum_{k_2} \left(\sum_{k_1} b[k_1, k_2] e^{-j\omega_1 k_1} \right) z_2^{k_2}.$$

To ensure continuity, the phase function $\phi(\omega_1, \omega_2)$ is defined as

$$\phi(\omega_1, \omega_2) = \text{Im} \left[\int \frac{\partial B_{\omega_1}(z_2)}{\partial z_2} [B_{\omega_1}(z_2)]^{-1} dz_2 \right] + \phi(\omega_1, 0), \qquad (4.20)$$

where the contour integral starts at $z_2 = 1$ and traverses the unit circle to $z_2 = e^{-j\omega_2}$. $\phi(\omega_1, \omega_2)$ is referred to as the unwrapped phase, with proper choice of $\phi(\omega_1, 0)$ as in (4.21) below.

The following theorem linking the phase function to stability of a 2-D filter has been found to be useful in the construction of efficient numerical tests.

Theorem 4.5. *The bivariate polynomial $B(z_1, z_2) \neq 0$ in $\overline{U}^2$ if and only if the unwrapped phase is continuous, odd, and periodic.*

Proof. If the phase is continuous, odd, and periodic, then $B(z_1, z_2) \neq 0$ on T^2 follows from the definition of $\phi(\omega_1, \omega_2)$ in (4.20). Moreover, as a specialization, the phase of $B(z_1^{k_1}, z_1^{k_2})$, where k_1, k_2 are nonnegative integers, must be continuous, odd, and periodic, implying in particular that $B(z_1, 1)$ and $B(1, z_2)$ have continuous, odd, and periodic phases. Invoking Fact 4.8, then, $B(z_1, z_2) \neq 0$ in $\overline{U}^2$.

When $B(z_1, z_2) \neq 0$ in $\overline{U}^2$, $\phi(\omega_1, \omega_2)$ can be defined as in (4.20). The constant $\phi(\omega_1, 0)$ is defined as follows:

$$\phi(\omega_1, 0) = \left[\int_0^{\omega_1} \frac{(\partial B_I/\partial \omega_1) B_R - (\partial B_R/\partial \omega_1) B_I}{B_R^2 + B_I^2} d\omega_1 \right]_{\omega_2=0} \qquad (4.21)$$

where B_R, B_I represent, respectively, the real and imaginary parts of $B(e^{-j\omega_1}, e^{-j\omega_2})$. With $\phi(\omega_1, 0)$ chosen as in (4.21), $\phi(\omega_1, \omega_2)$ in (4.20) is guaranteed to be continuous and odd. So $\phi(\omega_1, \omega_2) = -\phi(\omega_1, \omega_2)$. Using elementary arguments from complex variable theory, it can be shown, furthermore, that

$$\phi(\omega_1, \omega_2 + 2\pi) = \phi(\omega_1, \omega_2) + 2\pi k_{\omega_2}$$

$$\phi(\omega_1 + 2\pi, \omega_2) = \phi(\omega_1, \omega_2) + 2\pi k_{\omega_1},$$

where $k_{\omega_1}, k_{\omega_2}$ are, respectively, independent of ω_2, ω_1 and are actually constants. Then

$$\phi_a(\omega_1, \omega_2) = \phi(\omega_1, \omega_2) - k_{\omega_2}\omega_2 - k_{\omega_1}\omega_1,$$

where the phase function $\phi_a(\omega_1, \omega_2)$ associated with the finite-extent shifted array $\{b[k_1 + k_{\omega_1}, k_2 + k_{\omega_2}]\}$ is continuous, odd, and periodic. The proof of the theorem is now complete, after the observation that the polynomial $B(z_1, z_2)$ associated with $\{b[k_1, k_2]\}$ is nonzero in $\overline{U}^2$ (see Theorem 4.1).

The tests for zeros of $B(z_1, z_2) \neq 0$ in $\overline{U}^2$ which involve the test for zeros of $B(z_1, z_2)$ in T^2 along with other single-variable tests are all special cases of a theorem proved by Rudin using homotopy theory, and summarized next. See reference for a simple proof of this result.

Theorem 4.6. *The bivariate polynomial $B(z_1, z_2) \neq 0$ in $\overline{U}^2$ if and only if:*

(i) $B[f_1(z_1), f_2(z_1)] \neq 0$ in $|z_1| \leq 1$, where $f_1(z_1)$ and $f_2(z_1)$ are some continuous (not necessarily holomorphic) mappings of $\overline{U}^1$ into $\overline{U}^1$ and of T^1 into T^1 with positive winding number of the unit circle with respect to the origin, i.e., $\mathrm{Ind}f_1(e^{i\theta}) > 0$, $\mathrm{Ind}f_2(e^{i\theta}) > 0$, $0 \leq \theta \leq 2\pi$. ("Ind" denotes "index" or "winding number".)
(ii) $B(z_1, z_2) \neq 0$ in T^2.

This theorem is actually independent of the number n of variables and applies not only to polynomials but also to the class of all continuous complex functions on the closure $\overline{U}^n$ of U^n whose restriction to U^n is holomorphic. The theorem can be applied to generate other tests, almost at will. Of course, such tests may not be computationally efficient for actual implementation. One such test is stated next.

Fact 4.9. *The bivariate polynomial $B(z_1, z_2) \neq 0$ in $\overline{U}^2$ if and only if*

$$B(z_1, z_1 e^{j\alpha}) \neq 0 \quad in \quad \overline{U}^1, \quad for\ all\ \alpha\ in\ 0 \leq \alpha \leq 2\pi.$$

It may be noted that a result paralleling the one just stated applies in the test for zeros of $B(z_1, z_2)$ in U^2:

Fact 4.10. *The bivariate polynomial $B(z_1, z_2) \neq 0$ in U^2 if and only if*

$$B(z_1, z_1 e^{j\alpha}) \neq 0 \quad in \quad U^1, \quad for\ all\ \alpha\ in\ 0 \leq \alpha \leq 2\pi$$

A LSI n-dimensional recursive filter is characterizable by a transfer function,

$$H(z_1, z_2, \cdots, z_n) = \frac{A(z_1, z_2, \cdots, z_n)}{B(z_1, z_2, \cdots, z_n)}$$

where $H(z_1, z_2, \cdots, z_n)$ is viewed as a rational function of $z_1, z_2, \cdots, z_n, z_1^{-1}, z_2^{-1}, \cdots,$ z_n^{-1} (recursible filters include causal and weakly causal filters). For notational convenience, the z-transform $H(z_1, z_2, \cdots, z_n)$ of an n-D sequence, $\{h[k_1, k_2, \cdots, k_n]\}$, will be defined as a power series involving the superposition of the products of monomials of the type $z_1^{k_1}, z_2^{k_2}, \cdots, z_n^{k_n}$ and the generic element, $h[k_1, k_2, \cdots, k_n]$, of the sequence. Physically, the indeterminates $z_1, z_2, \cdots, z_n$ are the respective delay variables along the spatial or temporal directions of sampling during the analog to digital conversion of a multidimensional spatio-temporal signal. In the case of first quadrant quarter-plane filters, $A(z_1, z_2, \cdots, z_n)$ and $B(z_1, z_2, \cdots, z_n)$ are polynomials. Since a ring isomorphism maps a weakly causal filter onto a first quadrant quarter-plane filter, $A(z_1, z_2, \cdots, z_n)$, $B(z_1, z_2, \cdots, z_n)$, will be understood to be relatively prime polynomials in the delay variables $z_1, z_2, \cdots, z_n$ unless mentioned otherwise.

For a first quadrant quarter-plane digital filter, characterized by $H(z_1, z_2, \cdots, z_n)$, which is assumed to be holomorphic around the origin (this is assured by assuming $B(0, 0, \cdots, 0) \neq 0$), thereby permitting a Taylor series expansion,

$$H(z_1, z_2, \cdots, z_n) = \sum_{k_1=0}^{\infty} \cdots \sum_{k_n=0}^{\infty} h[k_1, \cdots, k_n] z_1^{k_1} \cdots z_n^{k_n} \qquad (4.22)$$

the investigation into BIBO stability reduces to the determining of conditions under which the sum

$$\sum_{k_1=0}^{\infty} \cdots \sum_{k_n=0}^{\infty} |h[k_1, \cdots, k_n]| \qquad (4.23)$$

converges. Absolute convergence implies that $h[k_1, \cdots, k_n]$'s are uniformly bounded, i.e. there exists a constant K such that, $|h[k_1, \cdots, k_n]| \leq K, k_1, \cdots, k_n = 0, 1, 2, \cdots$. It is well-known that convergence of (4.23) implies uniform convergence in $\overline{U}^n$ of (4.22) which in turn implies that $H(z_1, \cdots, z_n)$ is holomorphic in U^n and continuous in $\overline{U}^n$. Also, if $H(z_1, \cdots, z_n)$ is holomorphic in a neighborhood of $\overline{U}^n$, then (4.23) converges. In the $n = 1$ case, for a rational function, $H(z_1) = [A(z_1)]/[B(z_1)]$, it is simple to establish that (4.23) with $n = 1$ is absolutely summable if and only if the polynomial $B(z_1) \neq 0, |z_1| \leq 1$. This fact does not generalize in the $n > 1$ case and $B(z_1, \cdots, z_n) \neq 0$ in $\overline{U}^n$ is only a sufficient condition for BIBO stability of $H(z_1, \cdots, z_n)$.

For an arbitrary rational function $H(\mathbf{z})$, the problem of determining conditions for BIBO stability in the presence of a finite number of nonessential singularities of the second kind on T^n is complex and, at present, no general solution is available.

Dautov clarified the problem in the 2-D case by showing that for a certain class of denominator polynomials $B(\mathbf{z})$, $H(\mathbf{z}) = A(\mathbf{z})/B(\mathbf{z})$ is BIBO stable if and only if it can be continuously extended to $\overline{U}^2$ from U^2; furthermore, he conjectured that this is true for any $B(\mathbf{z})$ whose only zeros in $\overline{U}^2$ are on T^2 and these are also zeros of $A(\mathbf{z})$ (note that in the 2-D case the finiteness condition for the number of common zeros on T^n is automatically satisfied). Of course, $H(\mathbf{z})$ can never be extended analytically to $\overline{U}^n$ when there is a nonessential singularity of the second kind at $\mathbf{z} = \mathbf{z}^{(0)} \in T^n$ since in any open neighborhood of $\mathbf{z} = \mathbf{z}^{(0)}$, $H(\mathbf{z})$ will always be unbounded and therefore Riemann's analytic continuation theorem will not apply. However, if $H(\mathbf{z})$ extends continuously to $\overline{U}^n$ one can take limits from any direction from within $\overline{U}^n$ to achieve the same result. The example given next shows a rational function $H(\mathbf{z})$ which is analytic in U^2, has a nonessential singularity of the second kind at $\mathbf{z} = (1, 1)$ and does not extend on $\overline{U}^2$ to a continuous function.

Example 4.7. Let,

$$H(z_1, z_2) = \frac{(z_1 - 1)^n + (z_1 - 1)(z_2 - 1) + (z_2 - 1)^n}{(z_1 + z_2 - 2)^2}$$

$n \gg 1$. For $(z_1, z_2) = (e^{j\theta}, e^{j\phi}) \in T^2$,

$$H(e^{j\theta}, e^{j\phi}) = \frac{(e^{j\theta} - 1)^n + (e^{j\theta} - 1)(e^{j\phi} - 1) + (e^{j\phi} - 1)^n}{(e^{j\theta} - 1 + e^{j\phi} - 1)^2}$$

Then, the limit as $\theta \to 0$ along the line $\phi = 0$ is 0, but the limit as $\theta \to 0$ along the line $\theta = \phi$ is 1/4.

Through a series of neatly constructed examples, Goodman, in a prize-winning paper [64], was the first to point out difficulties in the prevailing concept of BIBO stability for multidimensional filters. In addition to other results, he showed that

$$H(z_1, z_2) = \frac{(1 - z_1)^m (1 - z_2)^n}{2 - z_1 - z_2}, \quad m \geq 0, \quad n \geq 0 \tag{4.24}$$

is BIBO unstable when $m = n = 1$, and BIBO stable when $m = n = 8$. For the case when $m = n = 1$ it is clear that $H(z_1, z_2)$ does not extend on $\overline{U}^2$ to a continuous function, since

$$\lim_{z_1 \to 1} H(z_1, z_1) = \lim_{z_1 \to 1} \frac{(1 - z_1)^2}{2(1 - z_1)} = 0$$

while

$$\lim_{x \to 1} H(1 - x^2 + jx, 1 - x^2 - jx) = \lim_{x \to 0} \frac{x^4 + x^2}{x^2} = 1.$$

For 1-D filters characterized by a rational transfer function $H(z_1)$ with a power series expansion about $z_1 = 0$ given by

$$H(z_1) = \sum_{k_1=0}^{\infty} h[k_1] z_1^{k_1},$$

the impulse response sequence $\{h[k_1]\}$ is known to satisfy the following properties in order that the filter might be BIBO stable:

(a)

$$\lim_{k_1 \to \infty} |h[k_1]| = 0$$

(b)

$$\sum_{k_1=0}^{\infty} |h[k_1]|^p < \infty, \quad \text{for any} \quad p \geq 1$$

(c)

$$\lim_{k_1 \to \infty} \sup(|h[k_1]|)^{1/k_1} < 1$$

(d)

$$|h[k_1]| \leq cr^{k_1}, \quad 0 \leq c < \infty, \quad 0 < r < 1.$$

These conditions are all equivalent, and each is necessary and sufficient for BIBO stability of the filter.

Consider, now a 2-D first quadrant filter characterized by a rational transfer function, with the following power series expansion about $z_1 = 0, z_2 = 0$:

$$H(z_1, z_2) = \frac{A(z_1, z_2)}{B(z_1, z_2)} = \sum_{k_1=0}^{\infty} \sum_{k_2=0}^{\infty} h[k_1, k_2] z_1^{k_1} z_2^{k_2}$$

It is assumed that $B(0,0) \neq 0$ and $A(z_1, z_2)$ and $B(z_1, z_2)$ are relatively prime polynomials. It will be shown that significant differences occur here from the 1-D case. It has been proved that the unstable filter with transfer function

$$H(z_1, z_2) = \frac{(1 - z_1)(1 - z_2)}{2 - z_1 - z_2} \tag{4.25}$$

having nonessential singularities of the second kind at $z_1 = z_2 = 1$ has a square summable impulse response, i.e., $\{h[k_1, k_2] \in l_2$, and therefore, also,

$$\lim_{k_1, k_2 \to \infty} h[k_1, k_2] = 0 \tag{4.26}$$

It has also been shown that the unstable filter transfer function

$$H(z_1, z_2) = \frac{2}{2 - z_1 - z_2} \qquad (4.27)$$

devoid of nonessential singularities of the second kind on T^2, has an impulse response which trails off of zero as in (4.26). However, the impulse response associated with $H(z_1, z_2)$ in (4.27) is not square summable. To determine the class of 2-D filter transfer functions for which counterparts of 1-D properties hold, assume that the rational transfer function $H(z_1, z_2)$ is devoid of nonessential singularities of the second kind on T^2.

Theorem 4.7. *Consider L to be defined as follows:*

$$L = \lim_{k_1, k_2 \to \infty} \sup |h[k_1, k_2]|^{1/(k_1 + k_2)} \qquad (4.28)$$

where $h[k_1, k_2]$'s are the coefficients of the power series expansion about $z_1 = 0$, $z_2 = 0$ of $H(z_1, z_2)$. Then the filter characterized by $H(z_1, z_2)$ is BIBO stable if and only if $L < 1$.

Proof. "If" part. From the definition of L, for any $\epsilon, 0 < \epsilon < 1, |h[k_1, k_2]| \leq [L(1 + \epsilon)]^{k_1 + k_2}$ for all but a finite number of pairs (k_1, k_2). If $L < 1$, it is possible to choose ϵ such that $L(1 + \epsilon) < 1$. Consequently, $\sum_{k_1} \sum_{k_2} |h[k_1, k_2]| < \infty$ and the filter is BIBO stable.

"Only if" part. Since the filter is BIBO stable, it follows from Fact 4.1 that $\sum_{k_1} \sum_{k_2} |h[k_1, k_2]| < \infty$. It is known that for an absolutely convergent double series, the limit

$$W_n(\phi)_{n \to \infty} \triangleq \sum_{k_1} \sum_{k_2} |h[k_1, k_2]|, \quad \phi[k_1, k_2] < n$$

exists for any admissible function $\phi[k_1, k_2]$ acting on the indices k_1, k_2 in $\{h[k_1, k_2]\}$ (for example, $\phi[k_1, k_2]$ could be $k_1 + k_2$), and this limit is independent of $\phi[k_1, k_2]$. Therefore, it is possible to infer using standard results for 1-D sequences that

$$\lim_{k_1, k_2 \to \infty} \sup |h[k_1, k_2]|^{1/(k_1 + k_2)} < 1,$$

and the proof of the theorem is complete.

From Theorem 4.7 it also follows that

$$|h[k_1, k_2]| \leq cr^{k_1 + k_2}, \quad 0 \leq c < \infty, \quad 0 < r < 1.$$

when $L < 1$. The case when $L = 1$ leads to instability and the locations of the unstable singularities are restricted by the assertion made next.

Fact 4.11. *In* (4.28), *if* $L = 1$, *the rational function* $H(z_1, z_2)$ *characterizes an unstable filter and the sources of instability may only occur in one of the following three regions:*

(i)

$$|z_1| = 1, \qquad z_2 \text{ arbitrary} \tag{4.29}$$

(ii)

$$|z_2| = 1, \qquad z_1 \text{ arbitrary} \tag{4.30}$$

(iii) At points $z_1 = z_{10}, z_2 = z_{20}$ *where*

$$H(z_{10}, z_{20}) = \infty, \quad |z_{10}| = 1, \quad |z_{20}| = 1. \tag{4.31}$$

The fact that if $L = 1$, then no unstable singularities of $H(z_1, z_2)$ occurs in U^2 can be substantiated as follows. The results of Theorem 4.7 suggest that if $L = 1, H(z_1, z_2)$ has unstable singularities in $|z_1| \leq 1, |z_2| \leq 1$. Let $z_1 = z_1' r, z_2 = z_2' r, 0 < r < 1$. Then

$$H(z_1, z_2) = H(rz_1', rz_2') = \sum_{k_1=0}^{\infty} \sum_{k_2=0}^{\infty} h[k_1, k_2] r^{k_1+k_2} (z_1')^{k_1} (z_2')^{k_2}$$

clearly,

$$\lim_{k_1, k_2 \to \infty} \sup(|h[k_1, k_2]| r^{k_1+k_2})^{1/(k_1+k_2)} = rL < 1.$$

Therefore, from Theorem 4.7, $H(z_1, z_2)$ has no unstable singularities in the open bidisc U^2. The proof for restriction of the unstable singularities to the regions described in (4.29)–(4.31) can be completed by the reader. The following theorem satisfies the objectives sought for exposition in this section.

Theorem 4.8. *If the rational function* $H(z_1, z_2)$ *does not belong to the class satisfying* (4.31), *then the following conditions are equivalent and each is necessary and sufficient for BIBO stability of the filter characterized by*

$$H(z_1, z_2) = \sum_{k_1=0}^{\infty} \sum_{k_2=0}^{\infty} h[k_1, k_2] z_1^{k_1} z_2^{k_2}$$

(i)

$$|h[k_1, k_2]| \leq c r^{k_1+k_2}, \quad 0 \leq c < \infty, \quad |r| < 1$$

(ii)

$$\sum_{k_1=0}^{\infty} \sum_{k_2=0}^{\infty} |h[k_1, k_2]|^p < \infty, \quad \text{for any} \quad p \geq 1$$

(iii)

$$\lim_{k_1,k_2 \to \infty} |h[k_1, k_2]| \to 0$$

(iv)

$$\lim_{k_1,k_2 \to \infty} \sup(|h[k_1, k_2]|)^{1/(k_1+k_2)} < 1.$$

It is recalled that $H(z_1, z_2)$ has been assumed to be devoid of nonessential singularities of the second kind on T^2. Furthermore, conditions (i) and (iv) are each necessary and sufficient for BIBO stability in general, though conditions (i) and (iii) each carry different implication from its 1-D counterpart.

This result has a direct extension to the $n > 2$ case. For a rational function $H(z_1, \cdots, z_n)$ to be outside the class satisfying the n-variate counterpart of (4.29)–(4.31), it must be devoid of nonessential singularities of the first kind on T^n.

Several further comments and observations pertaining to the $n = 2$ (or $n > 2$) case can be made. Though a bounded impulse response, i.e., $|h[k_1, k_2]| < K < \infty, \forall\ k_1, k_2$, does not imply BIBO stability of $H(z_1, z_2) = A(z_1, z_2)/B(z_1, z_2)$, it does imply that $B(z_1, z_2) \neq 0$ in U^2. This can be substantiated as follows. Since $|h[k_1, k_2]| < K < \infty\ \forall\ k_1, k_2$, the power series $\sum_{k_1=0}^{\infty} \sum_{k_2=0}^{\infty} h[k_1, k_2] z_1^{k_1} z_2^{k_2}$ is absolutely convergent in U^2, implying that $H(z_1, z_2)$ has no nonessential singularity of the first kind in U^2 (and therefore no nonessential singularity of the second kind in U^2). Hence $B(z_1, z_2) \neq 0$ in U^2. However, the converse if false. For example, in

$$H(z_1, z_2) = 1/(1 - z_1 z_2)^2,$$

$B(z_1, z_2) \neq 0$ in U^2. However,

$$h[k_1, k_2] = \begin{cases} 0, & \text{if}\ k_1 \neq k_2 \\ k_1 + 1, & \text{if}\ k_1 = k_2 \end{cases}$$

becomes unbounded as k_1 and k_2 approach infinity.

Some of the results discussed in this section are summarized in Table 4.1. The reader is advised to try to prove the last three statements in the table. For brevity, the similarities and differences from the univariate case have been identified via detailed exposition of the $n = 2$ case. Naturally, the main conclusions in Table 4.1 apply for the $n > 2$ case also.

Table 4.1 Summary of Algebraic 2-D Stability Tests

$B(z_1, z_2) \neq 0 \quad \text{in} \quad \overline{U}^2$	$\dfrac{\Leftarrow \, /}{\Rightarrow}$	BIBO stability
$B(z_1, z_2) \neq 0 \quad \text{in} \quad \overline{U}^2 - T^2$	$\dfrac{\Leftarrow}{\Rightarrow \, /}$	BIBO stability
$\{h[k_1, k_2]\} \in l_1$	$\Leftrightarrow$	BIBO stability
$\{h[k_1, k_2]\} \in l_2$	$\dfrac{\Leftarrow}{\Rightarrow \, /}$	BIBO stability
$\lim\limits_{k_1, k_2 \to \infty} h[k_1, k_2] \to 0$	$\dfrac{\Leftarrow}{\Rightarrow \, /}$	BIBO stability
$B(z_1, z_2) \neq 0 \quad \text{in} \quad U^2$	$\dfrac{\Leftarrow}{\Rightarrow \, /}$	$|h[k_1, k_2]| \leq K < \infty, \quad \forall k_1, k_2$
$|H(z_1, z_2)| \leq K < \infty \text{ in } U^2$	$\dfrac{\Leftarrow \, /}{\Rightarrow}$	$\{h[k_1, k_2]\} \in l_2$
$B(z_1, 0) \neq 0, \quad |z_1| \leq 1$	$\dfrac{\Leftarrow \, /}{\Rightarrow}$	$\sum_{k_1=0}^{\infty} |h[k_1, k_2]| < \infty, \quad \forall k_2$
$B(0, z_2) \neq 0, \quad |z_2| \leq 1$	$\dfrac{\Leftarrow \, /}{\Rightarrow}$	$\sum_{k_2=0}^{\infty} |h[k_1, k_2]| < \infty, \quad \forall k_1$
$H(z_1, z_2) = \frac{A(z_1, z_2)}{B(z_1, z_2)}$	$=$	$\sum_{k_1=0}^{\infty} \sum_{k_2=0}^{\infty} h[k_1, k_2] z_1^{k_1} z_2^{k_2}$

4.6.1 Advanced Notes

The distinguished boundary (referred to also as the torus) T^n of the open unit polydisc U^n is a compact Abelian group under the operation of the component-wise multiplication and as such carries a Lebesgue measure $m_n(T^n) = (2\pi)^n$. The polydisc algebra $A(U^n)$ is the class of all continuous complex functions on the closure $\overline{U}^n$ of U^n whose restriction to U^n is holomorphic. Several BIBO stability conditions in multidimensional signal processing emerge as special cases of a theorem due to Rudin for functions belonging to the class $A(U^n)$ [13, Theorem 4.7.2, p. 87]. This theorem essentially states that every value assumed on U^n by a function belonging to $A(U^n)$ is already assumed on a small subset $K \cup T^n$ of $\overline{U}^n$, for $K = \phi(\overline{U})$ where $\phi = (\phi_1, \ldots, \phi_n)$ is a continuous map of $\overline{U}$ into $\overline{U}^n$, which carries T into T^n such that for a loop E in T the winding number $Ind(\phi_j o E) > 0$ for $j = 1, 2, \ldots, n$. Further simplifications on the subset K, which have appeared in the signal processing literature, also follow directly by choosing $\phi_j(\lambda) = \lambda, 1 \leq j \leq n$, or indirectly from the result that if a function $f \in A(U^n)$ has no zeros on T^n, $f_j(\lambda) = f(1, \ldots, 1, \lambda, 1, \ldots, 1)$ (λ in jth place) and k_j is the number of zeros of f_j in U, then $f(\lambda, \ldots, \lambda)$ has $k_1 + \ldots + k_n$ zeros in U.

4.7 General Recursive Filters

For the non-quarter plane digital filter whose transfer function is

$$H(z_1, z_2) = \frac{Y(z_1, z_2)}{X(z_1, z_2)} = \frac{1}{1 + 0.5z_1 z_2^{-1} + z_1^{-1} + z_2^{-1} + z_1^{-2} z_2}, \tag{4.32}$$

the unit impulse response $\{h[k_1, k_2]\}$ is recursively computable with zero-boundary conditions from the 2-D difference equation

$$h[k_1, k_2] = \delta[k_1, k_2] - \frac{1}{2} h[k_1 + 1, k_2 - 1] - h[k_1 - 1, k_2]$$
$$- h[k_1, k_2 - 1] - h[k_1 - 2, k_2 + 1].$$

The support, $S_{p,q,r,t} \triangleq supp\{h[k_1, k_2]\}$, of the impulse response is in the causality cone $H_{1,1,1,2}$ defined by the intersection of the two half-planes,

$$H_{1,1} : x_1 + x_2 \geq 0 \text{ and } H_{1,2} : x_1 + 2x_2 \geq 0.$$

It is important to note that if $\mathcal{Z}$ denotes the set of integers and $\mathbf{Z}^2 \triangleq \mathbf{Z} \times \mathbf{Z}$, then there exists an index map

$$\phi : H_{p,q,r,t} \cap \mathbf{Z}^2 \rightarrow Q_1 \cap \mathbf{Z}^2,$$

from the causality cone $H_{p,q,r,t}$ with support $S_{p,q,r,t}$ to the first quadrant quarter-plane, Q_1 with support $S_{1,0,0,1}$.

The map referred to is bijective i.e. isomorphic (one-to-one and onto) and is described by the relation,

$$[n_1, n_2] = \phi[k_1, k_2] = [pk_1 + rk_2, qk_1 + tk_2]. \tag{4.33}$$

The associated inverse map is

$$[k_1, k_2] = \phi^{-1}[n_1, n_2] = [pn_1 - rn_2, -qn_1 + pn_2]. \tag{4.34}$$

Clearly,

$$\phi[t, -q] = [pt - qr, qt - qt] = [1, 0]$$

$$\phi[-r, p] = [-pr + pr, -qr + pt] = [0, 1]$$

which implies that points $[t, -q]$ and $[-r, p]$ in $H_{p,q,r,t}$ are mapped to points $[1, 0]$ and $[0, 1]$, respectively, in Q_1.

Quarter plane transfer function obtained from the image of the map is

$$\hat{H}(z_1, z_2) \triangleq \sum_{n_1=0}^{\infty} \sum_{n_2=0}^{\infty} h[\phi^{-1}[n_1, n_2]] z_1^{-n_1} z_2^{-n_2}$$

where $\phi^{-1}[n_1, n_2]$ is defined in (4.34). Then

$$\hat{H}[z_1, z_2] = \sum_{n_1=0}^{\infty} \sum_{n_2=0}^{\infty} h[tn_1 - rn_2, -qn_1 + pn_2] z_1^{-n_1} z_2^{-n_2}$$

$$= \sum \sum_{supp\{h[k_1, k_2]\}} h[k_1, k_2] z_1^{-(pk_1 + rk_2)} z_2^{-(qk_1 + tk_2)}$$

$$= \sum \sum_{supp\{h[k_1, k_2]\}} h[k_1, k_2] (z_1^p z_2^q)^{-k_1} (z_1^r z_2^t)^{-k_2}$$

$$= H(z_1^p z_2^q, z_1^r z_2^t).$$

Notice that for notational convenience, the definition here of the z-transform involves positive powers of the indeterminates, $w_1 \triangleq z_1^{-1}$, $w_2 \triangleq z_2^{-1}$. Therefore,

$$\hat{H}[z_1, z_2] = \sum \sum_{supp\{h[k_1, k_2]\}} h[k_1, k_2] (z_1^p z_2^q)^{-k_1} (z_1^r z_2^t)^{-k_2}$$

$$= \sum \sum_{supp\{h[k_1, k_2]\}} h[k_1, k_2] (w_1^p w_2^q)^{k_1} (w_1^r w_2^t)^{k_2}.$$

The transfer function in the complex variables w_1, w_2 of the first quadrant quarter-plane filter obtained from $H(z_1, z_2)$ in (4.32) is

$$G[w_1, w_2] = \left. \frac{1}{1 + 0.5 z_1 z_2^{-1} + z_1^{-1} + z_2^{-1} + z_1^{-2} z_2} \right|_{\substack{z_1 = w_1^{-1} w_2^{-1} \\ z_2 = w_1^{-1} w_2^{-2}}}$$

$$= \frac{1}{1 + 0.5 w_2 + w_1 w_2 + w_1 w_2^2 + w_1}$$

Apply first-quadrant quarter-plane stability test on $G(w_1, w_2)$ i.e. check whether or not $A(w_1, w_2) \neq 0$, $|w_1| \leq 1$, $|w_2| \leq 1$, where $A(w_1, w_2) \triangleq 1 + 0.5 w_2 + w_1 + w_1 w_2 + w_1 w_2^2$. Clearly $A(-1, 0) = 0$, which implies that $G(w_1, w_2)$ is BIBO unstable. Consequently, $H(z_1, z_2)$ in (4.32) is also a BIBO unstable general support

recursive digital filter. Indeed, from (4.32) it is straightforward to show that along
the boundaries of the causality cone that provides the region of support of the filter,

$$h[-k, k] = (-\frac{1}{2})^k, \quad k \geq 0$$

$$h[2k, -k] = (-1)^k, \quad k \geq 0.$$

Subsequently, it is easy to show after some algebraic manipulation

$$h[-k, k+1] = -\frac{3}{4}(k+1)(-\frac{1}{2})^k, \quad k \geq -1$$

$$h[2k+1, -k] = \frac{1}{2}k(-1)^k, \quad k \geq 0.$$

$$h[-k, k+2] = \left[-\frac{15}{2} + \frac{27}{4}(3+k)\frac{9}{4}(3+k)^2 \right](-\frac{1}{2})^{k+3}, \quad k \geq -2$$

$$h[-k, k+3] = \left[-\frac{27}{2} + \frac{153}{8}(4+k) - \frac{27}{4}(4+k)^2 + \frac{9}{8}(4+k)^3 \right](-\frac{1}{2})^{k+4}$$
$$k \geq -3$$

etc. Better still, plot the impulse response $h[k_1, k_2]$ by setting up recursive computation on (4.32) and see how it builds up, especially in the first quadrant.

4.8 2-D Complex Cepstrum

$$\hat{x}[k_1, k_2] = \frac{1}{(2\pi)^2} \int_0^{2\pi} \int_0^{2\pi} \ln X(w_1, w_2) e^{j(w_1 k_1 + w_2 k_2)} dw_1 dw_2$$

$\hat{X}(w_1, w_2) = \ln X(w_1, w_2)$ must be continuous, differentiable, and doubly periodic
and the function and the function $\ln X(z_1, z_2)$ must be analytic in some region of
convergence $\hat{R}$, where $R \supset \hat{R} \supset T^2$, with R the Region Of Convergence (ROC) of
$X(z_1, z_2)$. The function,

$$\hat{X}(w_1, w_2) \triangleq \ln |X(w_1, w_2)| + j\phi(w_1, w_2),$$

where

$$X(w_1, w_2) = |X(w_1, w_2)| + e^{j\phi(w_1 + w_2)}$$

is continuous and differentiable if $\phi(w_1, w_2)$ is the unwrapped phase. It can be shown that, in general, this unwrapped phase function is the sum of a doubly period phase function and a function linear in w_1 and w_2.

For the sake of simplicity, but without loss of generality, we assume that $x[k_1, k_2]$ is of finite support, so that

$$X(w_1, w_2) = \sum_{k_1} \sum_{k_2} x[k_1, k_2] e^{-j(w_1 k_1, w_2 k_2)}$$

is a trigonometric polynomial. First, consider

$$A_{w_1}(z_2) = X(w_1, w_2)\Big|_{\substack{w_1 \text{ fixed} \\ e^{jw_2} \to z_2}} = \sum_{k_1} \sum_{k_2} x[k_1, k_2] z_2^{-k_2} e^{-jw_1 k_1}$$

as an univariate polynomial in z_2^{-1}. Viewed as a rational function in z_2, $A_{w_1}(z_2)$ will have poles inside $|z_2| \leq 1$ only at $z_2 = 0$. This may be removed by multiplication with an appropriate power N_2 of z_2 to form the polynomial in z_2,

$$C_{w_1}(z_2) = z_2^{N_2} A_{w_1}(z_2).$$

Now, let us define the phase function

$$\phi(w_1, w_2) = I_m \int_{z_2 = e^{j0} = 1}^{e^{jw_2}} \frac{A'_{w_1}(z_2)}{A_{w_1}(z_2)} dz_2 + \phi(w_1, 0)$$

where the univariate phase function $\phi(w_1, 0)$ is the phase as a function of w_1 for $w_2 = 0$. Letting,

$$X(w_1, 0) = X_R(w_1, 0) + j X_I(w_1, 0),$$

$$\phi(w_1, 0) = \int_0^{w_1} \left(\frac{\frac{\partial X_I}{\partial w_1} X_R - \frac{\partial X_R}{\partial w_1} X_I}{X_R^2 + X_I^2} \right)\Bigg|_{w_2 = 0} dw_1$$

By constructing $\phi(w_1, w_2)$ in this manner, we are assumed that $\phi(w_1, w_2)$ is continuous and odd($\phi(w_1, w_2) = \phi(-w_1, -w_2)$), and we can write

$$\phi(w_1, 2\pi) = I_m \{ \oint \frac{A'_{w_1}(z_2)}{A_{w_1}(z_2)} dz_2 \}$$

$$= I_m \{ \oint \left(-N_2 z_2^{-1} + \frac{C'_{w_1}(z_2)}{C_{w_1}(z_2)} \right) dz_2 \}$$

$$= -2\pi N_2 + 2\pi r_2.$$

The last equation follows from application of residue theorem where r_2 is the number of roots (including multiplicities) of $C_{w_1}(z_2)$ and, therefore, of $A_{w_1}(z_2)$ inside the unit circle, $|z_2| \leq 1$. If we let $k_{w_2} = r_2 - N_2$, then

$$\phi(w_1, w_2 + 2\pi) = \phi(w_1, w_2) + 2\pi k_{w_2}$$

Similarly, we can derive that

$$\phi(w_1 + 2\pi, w_2) = \phi(w_1, w_2) + 2\pi k_{w_1}$$

What remains to be shown is that k_{w_2} is not a function of w_1 and k_{w_1} is not a function of w_2. If we examine the roots of $C_{w_1}(z_2)$, which are also the roots of $A_{w_1}(z_2)$, as we continuously vary the parameter w_1 from 0 to 2π, we discover that the roots move in a continuous manner. Thus, for a root to move from inside to outside the unit circle (or vice versa), it must lie on the unit circle, $|z_2| = 1$, for some value of w_1. This, however, violates the hypothesis $X(w_1, w_2) \neq 0$. Therefore, given a continuous odd function $\phi(w_1, w_2)$ such that

$$\phi(w_1, w_2 + 2\pi) = \phi(w_1, w_2) + 2\pi k_{w_2}$$

$$\phi(w_1 + 2\pi, w_2) = \phi(w_1, w_2) + 2\pi k_{w_1}$$

$$\phi(w_1, w_2) = -\phi(-w_1, -w_2),$$

we can subtract the linear phase component

$$\phi_L(w_1, w_2) = k_{w_1} w_1 + k_{w_2} w_2$$

to give the remaining term

$$\phi_A(w_1, w_2) = \phi(w_1, w_2) - \phi_L(w_1, w_2)$$

Setting, $k_{w_2} \rightarrow K_2, k_{w_1} \rightarrow K_1$, the sequence

$$y[k_1, k_2] = x[k_1 - K_1, k_2 - K_2]$$

which is a shifted version of $x[k_1, k_2]$ has a Fourier transform,

$$Y(w_1, w_2) = X(w_1, w_2)e^{-j(w_1 K_1 + w_2 K_2)}$$

Consequently, the unwrapped phase of $Y(w_1, w_2)$ is $\phi(w_1, w_2) - w_1 K_1 - w_2 K_2$, is continuous and double period, and $Y(w_1, w_2)$ satisfies the conditions necessary to define the complex cepstrum $\hat{y}[n_1, n_2]$.

4.8.1 Example

It is required to calculate the cepstrum $\hat{h}[k_1, k_2]$ of the unit impulse response sequence of a 2-D digital filter whose transfer function is $H(z_1, z_2) = \frac{1}{1 - az_1^{-1} - bz_2^{-1}}$, $|a| + |b| < 1$. For the sake of brevity in notation set $z_1^{-1} \rightarrow w_1$, and $z_2^{-1} \rightarrow w_2$. $|a| + |b| < 1$ implies that $H(w_1, w_2) \triangleq \frac{1}{1 - aw_1 - bw_2}$ is analytic in the unit bidisc $\bar{U}^2 \triangleq |w_1| \leq 1, |w_2| \leq 1$.

$$\hat{h}[k_1, k_2] = Z^{-1} \ln Z\{h[k_1, k_2]\}$$

$$\hat{h}[k_1, k_2] = \frac{1}{(2\pi j)^2} \oint_{|w_1|=1} \oint_{|w_2|=1} \ln\left(\frac{1}{1 - aw_1 - bw_2}\right)$$
$$w_1^{-(1+k_1)} z_2^{-(1+k_2)} dw_1 dw_2$$

$$= \frac{1}{(2\pi j)} \oint_{|w_2|=1} \frac{1}{w_2^{k_2+1}}$$

$$\left[\frac{1}{(2\pi j)} \oint_{|w_1|=1} \ln\left(\frac{1}{1 - aw_1 - bw_2}\right) \frac{1}{w_1^{k_1+1}} dw_1 \right] dw_2$$

Let $1 - aw_1 - bw_2 = 0$, $|w_2| = 1$, $w_1 = \frac{1 - bw_2}{a}$. Since

$$1 < \frac{1 - |b|}{a} \qquad \text{and} \qquad |w_2| = 1,$$

$$|w_1| > \frac{1 - |bw_2|}{|a|} > 1.$$

$$\ln\left(\frac{1}{1 - aw_1 - bw_2}\right) \text{ is analytic in } \bar{U}^2,$$

$$\frac{1}{(2\pi j)} \oint_{|w_1|=1} \ln\left(\frac{1}{1 - aw_1 - bw_2}\right) \frac{1}{w_1^{k_1+1}} dw_1$$

$$= \frac{1}{k_1!} \frac{d^{k_1}}{dw_1^{k_1}} \left(\ln\left(\frac{1}{1 - aw_1 - bw_2}\right) \right) \Big|_{w_1=0}$$

$$= \frac{a^{k_1}}{k_1 (1 - bw_2)^{k_1}}$$

$$\hat{h}[k_1, k_2] = \frac{1}{(2\pi j)} \oint_{|w_2|=1} \frac{1}{w_2^{k_2+1}} \left(\frac{a^{k_1}}{k_1 (1 - bw_2)^{k_1}} \right) dw_2$$

Since $|b| < 1, 1 - bw_2 = 0 \Rightarrow |w_2| = |\frac{1}{b}| > 1$.

$$\hat{h}[k_1, k_2] = \frac{u^{k_1}}{k_1(k_2!)} \frac{d^{k_2}}{dw_2^{k_2}} \left(\frac{1}{(1 - bw_2)^{k_1}} \right) \Big|_{w_2 = 0}$$

$$= \frac{(k_1 + k_2 - 1)!}{k_1! k_2!} a^{k_1} b^{k_2}, \qquad k_1 \geq 0, k_2 \geq 0, \qquad k_1 + k_2 \neq 0$$

4.9 Phase Unwrapping

Consider the n-D sequence $\{x[\mathbf{k}]\}$ whose Fourier transform

$$X_F(\mathbf{w}) \overset{\Delta}{=} X(e^{jw_1}, \ldots, e^{jw_n}),$$

is obtained by evaluating the z-transform $X(z_1, \ldots, z_n)$ of $\{x[\mathbf{k}]\}$ on T^n.

Definition 4.5 (Unwrapped Phase). For an n-dimensional signal $\{x[\mathbf{k}]\} \leftrightarrow X_F(\mathbf{w})$ such that $X_F(\boldsymbol{\mu}) \neq 0$ for some $\boldsymbol{\mu} = (\mu_1, \ldots, \mu_n) \in [0, 2\pi]^n$, the unwrapped phase $\theta_X(\mathbf{w}^*)$ at $\mathbf{w}^* = (w_1^*, \ldots, w_n^*) \in [0, 2\pi]^n$ is defined as

$$\theta_X(\mathbf{w}^*) \overset{\Delta}{=} \theta_X(\boldsymbol{\mu}) + \sum_{k=1}^{n} \int_{\mu_k}^{w_k^*} Im\{ \frac{(X_F^{(k)}(w_k))'}{X_F^{(k)}(w_k)} \} dw_k$$

where functions $X_F^{(k)}(w_k)$ for $k + 1, \ldots, n$ are given by

$$X_F^{(1)}(w_1) \overset{\Delta}{=} X_F(w_1, \mu_2, \ldots, \mu_n)$$

$$X_F^{(k)}(w_k) \overset{\Delta}{=} X_F(w_1^*, \ldots, w_{k-1}^*, w_k, \mu_{k+1}, \ldots, \mu_n) \text{ for } k = 2, \ldots, n.$$

and $(X_F^{(k)}(w_k))'$ denotes their derivative. All integrals here are defined by Lebesgue integral.

The phase unwrapping computation of a n-D signal can be reduced the computating of integrals of certain univariate functions as is evident from the Lemma next.

Lemma 4.1. *Under the same assumptions of the preceding Definition, let*

$$\min_\ell \overset{\Delta}{=} \min\{n_\ell | \mathbf{k} = [k_1, k_2, \ldots, k_n] \in Supp(x)\} \text{ for } \ell = 1, 2, \ldots, n$$

$$Y(z_1, \ldots, z_n) \overset{\Delta}{=} z_1^{-\min_1} \ldots z_n^{-\min_n} X(z_1, \ldots, z_n)$$

Define $Y_F^{(k)}(w_k)(k = 1, \ldots, n)$ in the same way as $X_F^{(k)}(w_k)$ in the preceding Definition. Then the one dimensional signals $\{y^{(k)}(n_k)\} \leftrightarrow Y_F^{(k)}(w_k)(k = 1, \ldots, n)$ satisfy

$$Supp(y^{(k)}) \subset \{n_k \in \mathbf{Z} | 0 \leq n_k\}$$

$$|Supp(y^{(k)})| \leq |Supp(x)| < \infty$$

and

$$\theta_X(\mathbf{w}^*) = \theta_X(\boldsymbol{\mu}) + \sum_{k=1}^n \left\{ \min_k(w_k^* - \mu_k) + \int_0^{w_k^*} Im\left(\frac{(Y_F^{(k)}(w_k))'}{Y_F^{(k)}(w_k)}\right) dw_k \right.$$
$$\left. - \int_0^{\mu_k} Im\left(\frac{(Y_F^{(k)}(w_k))'}{Y_F^{(k)}(w_k)}\right) dw_k \right\}.$$

A new approach to the stability of multivariate polynomials, with respect to a unit polydisc, was advanced in [37] as a special case of an algebraic characterization of the exact multidimensional unwrapped phase. The proposed phase unwrapping algorithm was also applied to the classical problem concerned with the zero distribution, with respect to the unit circle, for an arbitrary complex coefficient univariate polynomial without encountering the plethora of singular cases. By applying the theory of Cauchy indices, a symbolic algebra based analytic expression was also provided for the unwrapped phase (and, consequently, zero distribution with respect to the imaginary axis) associated with any complex coefficient characteristic polynomial of a continuous-time system [38]. The proposed algorithms in [37] and [38] do not require any zero-finding and, very importantly, force the singular case problem in all division algorithm based procedures, to be absent.

4.10   Conclusions

While two-dimensional systems theory was understood well prior to 1985, the possibilities and intrinsic difficulties in n-D, $n \geq 3$ have been understood more fully during the last twenty years or so. This is largely due to the progress in n-D, $n \geq 3$ polynomial matrix theory and matrix fraction descriptions on one hand [6, 39, 65] and the behavioral approach on the other [66–68].

The zero set of a multivariate polynomial is unbounded and, often it is necessary to determine whether or not a specified polydomain (or its complement) is zero-free for a polynomial. In [69], the zero set of a multivariate polynomial is enclosed by unions and intersections of unbounded sets. Two additions to the extensive literature on implementation of 2-D stability tests are available in [70] and [71]. A member of the class $A(U^n)$ qualifies as the transfer function of a normal (first quadrant

BIBO-stable linear shift-invariant) filter if it has finite norm (defined as equivalent to the filter being BIBO-stable). The work of Dautov [72] suggested the conjecture that all rational functions in $A(U^n)$, including those that have nonessential singularities of the second kind (NSSK) on T^n, have finite norm. Youla [73] advanced a system-theory motivated proof of Rudin's result that every element belonging to the special class of all-pass functions in $A(U^n)$ is rational, devoid of NSSK on T^n and of finite norm. In a recent paper [74], it was proved that a rational multivariate first quadrant quarter-plane digital filter transfer function, analytic on the open polydisc, is BIBO-stable if and only if it has a uniform extension to the distinguished boundary of the polydisc.

Appendix: Editorial Note

Since the writing of the 1982 edition of the text, significant developments on characterization of stability of n-D ($n \geq 2$) systems, their tests, and robustness, including a variety of new applications of these concepts, have emerged. Since the literature on this topic is large we only point to key references on each of these developments.

A. The relevance of *BIBO stability* has been questioned on the basis of various considerations, and a set of stability criteria all of which coincide with BIBO stability in the 1-D case, but are different in n-D ($n \geq 2$) have been proposed. For example, passive circuit theoretic considerations gave rise to previously unidentified *classes of stable multidimensional polynomials* with zeros on the distinguished boundary. The so called scattering Hurwitz (Schur), reactance Hurwitz (Schur) and the immittance Hurwitz (Schur) polynomials, introduced in [A1] below, can all have different zero-structures on the distinguished boundary, and feature as the denominators (and numerators) of multidimensional scattering functions, reactance functions and immittance functions in pursuing generalizations of passive circuit theory to multidimension.[5] Properties of these classes of stable polynomials and their relevance to signals and systems are extensively discussed in and related other publications:

[A1] A. Fettweis and S. Basu, New results on multidimensional stable polynomials - Part I: Continuous case, *IEEE Trans. Circuits Syst.*, vol. 34, pp. 1221–1232, 1987.

---

[5]For terminological convenience we refer to polynomials devoid of zeros in the poly-disc, arising in discrete domain considerations, as the Schur polynomials, whereas the polynomials devoid of zeros in the poly-halfplane arising in continuous domain considerations as the Hurwitz polynomials. Correspondingly, we also talk about distinguished boundary of the poly-disc (denoted by T^n in the present text), or of the poly-halfplane.

[A2] S. Basu and A. Fettweis, New results on stable multivariable polynomials - Part II: Discrete case. *IEEE Trans. Circuits Syst.*, vol. 34, pp. 1257–1270, November 1987.

[A3] S. Basu, New results on stable multidimensional polynomials - Part III: State space interpretations. *IEEE Trans. Circuits Syst.*, vol. 38, no. 7, July 1991.

B. *Homotopy* arguments, as a technique borrowed from analytic function theory, have been mentioned for proofs of Theorems 4.4 and 4.6. However, it was shown later that by carefully using the elementary fact that zeros of a polynomial are continuous functions of coefficients of the polynomial, an entire gamut of equivalence results including Theorems 4.3–4.6 and Facts 4.5–4.9, and several yet further results not mentioned here, can indeed be proven. The success of this elementary yet powerful technique, while nascent in several references mentioned here, is most convincingly demonstrated in:

[B1] S. Basu and A. Fettweis, Simple proofs of some discrete domain stability related properties of multidimensional polynomials, *Int. J. Circuit Theory Applns.*, vol. 15, pp. 357–370, 1985.

C. Robustness of stability of polynomials is a topic that emerged from the original work of Russian scientist V. L. Kharitonov much later than 1982, and was subsequently investigated in the multidimensional context by several researchers, including N. K. Bose himself. Here, the invariance of the stability property of a polynomial (e.g., the Hurwitz or the Schur property mentioned earlier) under coefficient perturbation is the topic of interest. While the relevant literature is large, an early comprehensive treatment of the multidimensional case appears in:

[C1] S. Basu, On boundary implications of stability and positivity properties of multidimensional systems, *Proc. IEEE*, Special Issue on Multidimensional Signal Processing (Editor N. K. Bose), vol. 78, no. 4, pp. 614–626, May 1990.

In this vein, a more recent treatment of robust stability, from which the interested reader can delve deeper into the topic, is a publication by V. L. Kharitonov himself:

[C2] J. A. Torres-Munoz, E. Rodriguez-Angeles, and V. L. Kharitonov, On Schur Stable Multivariate Polynomials, *IEEE Trans. Circuits Syst.-I: Regular papers*, vol. 53, no. 5, May 2006.

D. As for algebraic *tests for stability* (cf. Sect. 4.6 of this chapter), while early work was motivated by demonstrating "decidability" of the problem, i.e., existence of an algorithm that terminates in a finite number of steps, little attention was paid to computational complexity of the algorithms involved(cf. reference [3.37] in the 1982 edition). Indeed, early algorithms were of exponential complexity, and subsequent considerations of complexity eventually led to a new generation of algorithms for 2-D stability test, the best known being the $O(n^4)$ algorithm due to Yuval Bistritz, the details of which are available e.g., in:

[D1] Y. Bistritz, Stability Testing of 2-D Discrete Linear Systems by Telepolation of an Immittance Type Tabular Test, *IEEE Trans. Circuits Syst.-I: Fundamental Theory Applns.*, vol. 48, no. 7, July 2001.

[D2] Y. Bistritz, Real polynomial based immittance type tabular stability test for two dimensional discrete systems, *Circuits Systems and Signal Processing*, vol. 22, no. 3, pp. 255–276, 2003, Birkhauser, Boston.

E. The surprising connection between stable multivariable polynomials arising in circuits and systems theory on the one hand and *statistical mechanics, network reliability* and the theory of *matroids* on the other, was first exploited in [E1] below and appeared in the definitive paper [E2, Theorem 3.4 and Lemma 3.5] by following results from [A1] above. Assuring reliability of a network with probabilistic edge weights can be formulated as a problem of examining the roots of a so called reliability polynomial in the unit poly-disc. In statistical mechanics, critical phenomena exhibited by a system described by, say, an Ising model is determined by zeros of the associate partition function, which could be a polynomial in several variables, and in the theory of matroids the set of bases for matroids of certain types can be associated with coefficients of multivariable (stable) Hurwitz polynomials. Since the publication of [E2] there has been considerable activity in statistical mechanics, and combinatorial graph algorithms along these lines, and it has been conjectured [E2, E3] that stability results on multivariable polynomials may be further useful for investigating a long list of open problems in the area.

[E1] E. H. Lieb and A. Sokal, A general Lee-Yang Theorem for one-component and multicomponent Ferromagnets, *Communications in Mathematical Physics*, vol. 80, pp. 153–179, Springer-Verlag, 1981.

[E2] Y.-B. Choe, J. G. Oxley, A. D. Sokal, and D. G. Wagner, Homogeneous multivariate polynomials with the half-plane property, *Advances in Applied Mathematics*, vol. 32, Issues 1–2, pp. 88–187, Jan.-Feb. 2004.

[E3] A. D. Sokal, The multivariable Tutte polynomial (alias Potts model) for graphs and matroids, *Surveys in Combinatorics*, Bridget S. Webb (ed.), London Mathematical Society Lecture Note Series, pp. 173–226. 2005.

Chapter 5
2-D FIR Filters, Linear Prediction and 2-D IIR Filters

5.1 Introduction

The technical field of image processing began during the latter half of the nineteenth century in the early days of chemical photography. Image processing then branched into the evolving fields of radio wave transmission and X-ray technology during the early twentieth century. Based on this evolutionary trend image processing became highly interdisciplinary during the twentieth century and has continued to quickly move forward during the twenty-first century. 2-D FIR filters are widely used in image processing because the feasibility of generating zero-phase or linear phase characteristics controls phase distortion due to the importance of phase in image processing.

5.2 Similarities and Differences with 1-D FIR Counterparts

In this section, feasibility of generalizations or lack of it is discussed and substantiated with respect to popular 1-D methods like window-based, frequency-sampling scheme, and the equiripple Parks-McClellan method [75] for designing FIR filters.

5.2.1 Windowing Scheme

Fact 5.1. *If $w(x)$ is a good symmetrical 1-D window, then $w\left(\sqrt{x_1^2 + x_2^2}\right)$ is a good circularly symmetrical 2-D window.*

© Springer International Publishing AG 2017   127
N.K. Bose, *Applied Multidimensional Systems Theory*,
DOI 10.1007/978-3-319-46825-9_5

The preceding fact has been used to design a good circularly symmetric 2-D window from a good symmetrical 1-D window [3, 76]. The approximation is especially good in 2-D FIR filter design if the support of the window transform is much smaller than the extent of band-limitedness of the low-pass 2-D circularly symmetric wavenumber response. In particular, let the ideal 2-D circularly symmetric wavenumber response be defined by

$$H\left(e^{j\omega_1}, e^{j\omega_2}\right) \triangleq H(\omega_1, \omega_2) = \begin{cases} 1, & \omega_1^2 + \omega_2^2 \leq B^2 \\ 0, & \text{otherwise} \end{cases}$$

Let the transform of the 2-D window which is used to truncate the 2-D impulse response, $h(x_1, x_2)$, be $W(\omega_1, \omega_2)$. Then, the actual wavenumber response is given by the 2-D convolution

$$F(\omega_1, \omega_2) = H(\omega_1, \omega_2) * *W(\omega_1, \omega_2)$$

Clearly, if the "width" of $W(\omega_1, \omega_2)$ is much smaller than the "width" of $H(\omega_1, \omega_2)$, $F(\omega_1, \omega_2)$ becomes essentially the convolution of $W\omega_1, \omega_2)$ and a 2-D impulse function. Therefore, if

$$w(x_1, x_2) = w\left(\sqrt{x_1^2 + x_2^2}\right)$$

then $\frac{1}{2\pi}F(\omega_1, \omega_2)$ is approximately of the functional form $F\left(\sqrt{\omega_1^2 + \omega_2^2}\right)$.

5.2.2   Frequency-Sampling Scheme

In 1-D, the Lagrange formula gives a polynomial $P(z)$ of degree n which interpolates over $(n + 1)$ distinct points $z_0, z_1, \ldots, z_n$. Specifically, if the frequency samples at $z_0, z_1, \ldots, z_n$ are specified by,

$$P(z_k) = w_k, \qquad k = 0, 1, \ldots, n, \qquad \text{then}$$

$$P(z) = \sum_{k=0}^{n} w_k \ell_k(z), \qquad \text{where}$$

$$\ell_k(z) = \frac{\prod_{i \neq k}(z - z_i)}{\prod_{i \neq k}(z_k - z_i)}$$

Thus, given $(n + 1)$ distinct points $z_0, z_1, \ldots, z_n$ and $(n + 1)$ values $w_0, w_1, \ldots, w_n$, there exists a unique polynomial $P_n(z)$ of degree n for which $P_n(z_i) = w_i$, $i = 0, 1, \ldots, n$.

In 2-D, a similar interpolation is possible. In particular, let $z_0, z_1, \ldots, z_n$ be $(n+1)$ distinct points. Let $w_0, w_1, \ldots, w_m$ be a second such set of $(m+1)$ points. Set

$$P(z) = \prod_{i=0}^{n}(z - z_i)$$

$$Q(w) = \prod_{j=0}^{m}(w - w_j)$$

$$P_j(z) \triangleq P(z)/(z - z_j)$$

$$Q_k(w) \triangleq P(w)/(w - w_k)$$

The $(m+1)(n+1)$ polynomials,

$$\ell_{jk}(z, w) \triangleq \frac{P_j(z)Q_k(w)}{P_j(z_j)Q_k(w_k)}$$

satisfy

$$\ell_{jk}(z_r, w_s) = \delta_{jr}\,\delta_{ks}$$

Hence,

$$P(z, w) = \sum_{j=0}^{n}\sum_{k=0}^{m}\mu_{jk}\ell_{jk}(z, w)$$

is a polynomial of degree $\leq m + n$ which satisfies the $(m+1)(n+1)$ interpolation conditions

$$P(z_j, w_k) = \mu_{jk}, \qquad j = 0, 1, \ldots, n, \qquad \text{and} \qquad k = 0, 1, \ldots, m$$

Thus, frequency sampling 2-D FIR filters can be designed.

5.2.3 Optimal Equiripple FIR Filters

The Park-McClellan method for optimal (in the minimax sense) 1-D FIR filter design is based on the validity of uniqueness of best approximation (Haar condition). This condition provides uniqueness of best approximation, a useful criterion (for convergence) in iterative implementation via the Remez exchange algorithm. The

Haar condition does not hold, in general, in the 2-D case. Namely, let the powers in two variables z, w be listed as follows: $P_0(z, w) = 1, P_1(z, w) = z, P_2(z, w) = w, P_3(z, w) = z^2, P_4(z, w) = zw, P_5(z, w) = w^2, P_6(z, w) = z^3, \ldots$. It is not always possible, having been given n arbitrary distinct points (x_j, w_j), to find a linear combination of $P_0, \ldots, P_{n-1}$ that takes on preassigned values at these points.

5.2.4 A Transformation Method for Design of Multidimensional FIR Filters

Let $H(e^{j\omega_1}, e^{j\omega_2}) \triangleq H(\omega_1, \omega_2)$ be the Fourier transform (FT) of the unit impulse response sequence $h[n_1, n_2]$. The zero-phase condition, up to signs, is described by

$$h[n_1, n_2] = h^*[-n_1, -n_2], \qquad H(\omega_1, \omega_2) = [H(\omega_1, \omega_2)]^*$$

A transformation method for design of zero-phase 2-D filters from a prototype 1-D zero-phase filter has been developed based on a 1-D to 2-D frequency transformation [77]. To reduce notational clutter, real-valued sequences are assumed. The FT of a real-valued sequence $\{h[n]\}_{n=-N}^{N}$, of length $2N + 1$, satisfying the even-symmetry condition, $h[n] = h[-n]$, for $n = 1, 2, \ldots, N$, is

$$H(e^{j\omega}) = H(\omega) = \sum_{k=0}^{N} a[k] \cos k\omega = \sum_{k=0}^{N} a[k] T_k(\cos \omega) \qquad (5.1)$$

where $a[0] \triangleq h[0], a[k] = 2h[k], k = 1, 2, \ldots, N$. The previous equation can be rewritten as

$$H(\omega) = \sum_{k=0}^{N} b[k] \cos^k \omega \qquad (5.2)$$

where the coefficients $b[k]$ are obtained from coefficients $a[k]$ by applying the inverse Chebyshev matrix transformation. To wit, for $N = 4$, and $x \triangleq \cos \omega$, the Chebyshev recursion gives

$$
\begin{bmatrix}
T_0(x) \\
T_1(x) \\
T_2(x) \\
T_3(x) \\
T_4(x)
\end{bmatrix}
=
\begin{bmatrix}
1 & 0 & 0 & 0 & 0 \\
0 & 1 & 0 & 0 & 0 \\
-1 & 0 & 2 & 0 & 0 \\
0 & -3 & 0 & 4 & 0 \\
1 & 0 & -8 & 0 & 8
\end{bmatrix}
\begin{bmatrix}
1 \\
x \\
x^2 \\
x^3 \\
x^4
\end{bmatrix}
$$

which following inversion of the lower-triangular matrix yields (for brevity, $T_k(x)$ is written below as T_k),

$$
\begin{bmatrix} 1 \\ x \\ x^2 \\ x^3 \\ x^4 \end{bmatrix} =
\begin{bmatrix}
1 & 0 & 0 & 0 & 0 \\
0 & 1 & 0 & 0 & 0 \\
\frac{1}{2} & 0 & \frac{1}{2} & 0 & 0 \\
0 & \frac{3}{4} & 0 & \frac{1}{4} & 0 \\
\frac{-3}{8} & 0 & \frac{1}{2} & 0 & \frac{1}{8}
\end{bmatrix}
\begin{bmatrix} T_0 \\ T_1 \\ T_2 \\ T_3 \\ T_4 \end{bmatrix} =
\begin{bmatrix} T_0 \\ T_1 \\ \frac{1}{2}(T_0 + T_1) \\ \frac{1}{4}(3T_1 + T_3) \\ \frac{1}{8}(3T_0 + 4T_2 + T_4) \end{bmatrix}
$$

To design a 2-D zero-phase filter from a 1-D prototype in (5.1), McClellan applied the transformation

$$
\cos\omega \triangleq F(\omega_1, \omega_2) = t[0,0] + t[1,0]\cos\omega_1 + t[0,1]\cos\omega_2 + t[1,1]\cos\omega_1\cos\omega_2.
\tag{5.3}
$$

For an arbitrary but fixed ω in the baseband, $-\pi \leqslant \omega \leqslant \pi$, consider the following function and derivative,

$$
\cos\omega_2 = \frac{\cos\omega - t[0,0] - t[1,0]\cos\omega_1}{t[0,1] + t[1,1]\cos\omega_1}
$$

$$
\frac{\partial(\cos\omega_2)}{\partial(\cos\omega_1)} = \frac{-t[0,1]t[1,0] + t[1,1]t[0,0] - t[1,1]\cos\omega}{(t[0,1] + t[1,1]\cos\omega_1)^2}
\tag{5.4}
$$

For fixed ω, $\cos\omega_2$ is either a monotone increasing or a monotone decreasing function of $\cos\omega_1$ because the sign of $\frac{\partial(\cos\omega_2)}{\partial(\cos\omega_1)}$ does not change as $\cos\omega_1$ varies from -1 to $+1$. The plot of $\cos\omega_2$ versus $\cos\omega_1$ and, therefore, of ω_2 versus ω_1 is either monotonically increasing or monotonically decreasing as a function of the parameter ω.

With properly chosen values of the parameters, the monotonically decreasing set of contours lead to circularly symmetric approximants. These approximants are better in the vicinity of origin and suitable for approximating low-pass circularly symmetric wavenumber characteristics. On the other hand, the monotonically increasing set of contours are suitable for the design of fan filters. The shapes of the contours in the two cases are shown in Fig. 5.1. McClellan calculated the set of parameters based on the conditions imposed in the transformation to satisfy the response characteristics of the filter types. For example, for the design of low-pass circularly symmetric filter response, it is necessary that $\omega = 0$ maps to $(\omega_1, \omega_2) = (0, 0)$, and $\omega = \pi$ maps to $(\omega_1, \omega_2) = (\pi, \pi)$. On imposition of these constraints to the transformation in (5.3), the following set of equations emerges.

$$
1 = t[0,0] + t[1,0] + t[0,1] + t[1,1]
$$

$$
-1 = t[0,0] - t[1,0] - t[0,1] + t[1,1]
$$

a

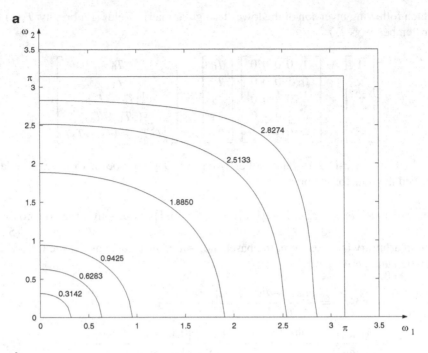

b

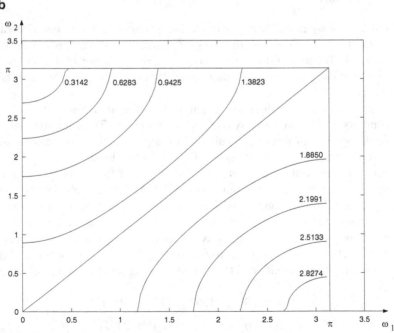

Fig. 5.1 Plot of monotonically decreasing and monotonically increasing contours that approximate (**a**) low-pass circularly symmetric filter characteristics and (**b**) fan filter characteristics

Therefore, $t[0,0] + t[1,1] = 0$ and $t[1,0] + t[0,1] = 1$. Note that from (5.4), $|t[1,1]| \leq |t[0,1]|$ and, interchanging the roles of ω_1, and ω_2, it then follows that $|t[1,1]| \leq |t[1,0]|$. An acceptable choice of parameters satisfying the constraints above is

$$t[0,1] = t[1,0] = t[1,1] = -t[0,0] = \frac{1}{2}$$

To design fan filters, the transformation in (5.1) is required to map $\omega = 0$ to $(\omega_1, \omega_2) = (0, \pi)$, and $\omega = \pi$ to $(\omega_1, \omega_2) = (\pi, 0)$. The resulting equations are

$$1 = t[0,0] + t[1,0] - t[0,1] - t[1,1]$$
$$-1 = t[0,0] - t[1,0] + t[0,1] - t[1,1]$$

An acceptable choice of parameters for the design of fan filters is

$$t[0,0] = t[1,1] = 0, \qquad t[1,0] = -t[0,1] = \frac{1}{2}$$

Note that the transformation in (5.1) must satisfy the condition

$$-1 \leq F(\omega_1, \omega_2) \leq 1, \qquad -\pi \leq \omega_i \leq \pi, \quad i = 1,2 \tag{5.5}$$

The plot $F(\omega_1, \omega_2) = K$, where K is a constant, generates a locus of points in the (ω_1, ω_2)-plane called isopotentials. A contour plot is defined to be several of these isopotentials. A contour plot of $F(\omega_1, \omega_2)$ looks identical in shape to the contour plot of the 2-D wavenumber response

$$H(e^{j\omega_1}, e^{j\omega_2}) \triangleq H(\omega_1, \omega_2) = \sum_{k=0}^{N} a[k]T_k[F(\omega_1, \omega_2)] \tag{5.6}$$

The value of $H(\omega_1, \omega_2)$ on a particular isopotential depends on both $F(\omega_1, \omega_2)$ (i.e. the parameters $t[0,0], t[1,0], t[0,1], t[1,1]$) and $a[k]$'s, the 1-D filter coefficients. Therefore, shape depends on the parameters and value depends on both parameters and the 1-D filter unit impulse response. Note that if a $F(\omega_1, \omega_2)$ does not satisfy the restriction in (5.3), then the modified one given by,

$$\hat{F}(\omega_1, \omega_2) = \frac{2}{F_{max} - F_{min}} F(\omega_1, \omega_2) - \frac{F_{max} + F_{min}}{F_{max} - F_{min}} \tag{5.7}$$

does and it has the same isopotential as $F(\omega_1, \omega_2)$.

The 2-D and higher dimensional generalizations of the original McClellan transformation have appeared in various forms. A 2-D generalization of $F(\omega_1, \omega_2)$ in (5.3) is

$$F(\omega_1, \omega_2) = \sum_{m=0}^{M} \sum_{n=0}^{N} t[m,n] \cos m\omega_1 \cos n\omega_2 \tag{5.8}$$

which was extensively studied by Mecklenbrauker and others, especially with respect to hardware implementation [5.3]. The hardware realization is facilitated by the first kind Chebyshev polynomial recursion

$$T_k[F(\omega_1, \omega_2)] = 2F(\omega_1, \omega_2)T_{k-1}[F(\omega_1, \omega_2)] - T_{k-2}[F(\omega_1, \omega_2)]$$

The 2-D filter as described by

$$H(e^{j\omega_1}, e^{j\omega_2}) \triangleq H(\omega_1, \omega_2) = h[0] + \sum_{k=1}^{N} 2h[k]T_k[F(\omega_1, \omega_2)]$$

is obtained after replacing $\cos \omega$ by $F(\omega_1, \omega_2)$ in the prototype 1-D frequency response,

$$H(e^{j\omega}) \triangleq H(\omega) = h[0] + \sum_{k=1}^{N} 2h[k]T_k(\cos \omega)$$

The modular realization, suitable for implementation in VLSI is shown in Fig. 5.2 when $N = 7$. This structure can be expanded or reduced very easily and the generic block $F(z_1, z_2)$ is realized from the transformation parameters while the transversal taps are set by the 1-D FIR filter. It is also noteworthy that the McClellan transformation that maps a 1-D prototype to a computationally efficient 2D FIR filter has also been used to realize computationally efficient adaptive 2D FIR digital filters where the adaptive computations are based on updating the tap weights of the 1-D prototype [78].

Dudgeon used a 3-D generalization of the form

$$F(\omega_1, \omega_2, \omega_3) = \frac{1}{4}(\cos \omega_1 + \cos \omega_2 + \cos \omega_3 +$$

$$\cos \omega_1 \cos \omega_2 + \cos \omega_1 \cos \omega_3 + \cos \omega_2 \cos \omega_3 +$$

$$\cos \omega_1 \cos \omega_2 \cos \omega_3 - 3)$$

in the use of multidimensional FIR filter design based on McClellan-type transformations for beam-forming and digital array processing.

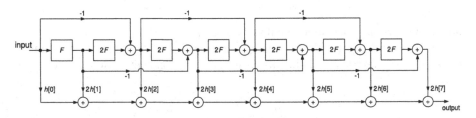

Fig. 5.2 The generic form of modular realization for the $N = 7$ case

5.3 Fan, Directional, and Velocity Filters

A fan filter is a 2-D filter whose passband region is limited by two straight boundary lines passing through the origin that can point to arbitrary direction. 2-D signals often exhibit certain orientation in space called directions. When a particular signal contain components with a given direction, its Fourier transform also contain high amplitude components in a well defined direction. This property can be exploited by the fan filter to either pass or reject a signal depending on whether its wavenumber spectrum is along the pass-direction of the filter. The concept of velocity filtering arose from array processing of seismic signals and the underlying principles are common to those in directional filtering except with a different interpretation for the sequences under consideration. Again, velocity filtering can be implemented with fan filters as discussed next.

Velocity filters are essentially unique in seismology (velocity-tuned spatial filters, have, however been designed to detect motion in an image sequence by using the phase information from the output of such filters) since the velocity of propagation is not constant for all waves [79, 80]. In radar, radio astronomy and to a large extent also in sonar, the velocity is essentially constant. Velocity filters provide a means to discriminate signals form noise or other undesired signals due to their different apparent velocities. Thus, the use of velocity filters makes it possible to enhance even signals occupying the same frequency ranges as those due to noise or undesired signals. Subsurface ground formation can be explored by the seismic reflection method. A seismic wave generated by an explosion of dynamite near the surface travels through different paths to different sedimentation layers and, subsequently, returns to the surface after reflection by the interfaces. A linear or rectangular array of sensors at the surface records the reflected seismic traces. The seismic waveforms consist of two distinct components: a component due to reflections from the subsurface formations and a surface component, usually referred to as ground roll. The required information is embedded in the reflected component. The ground roll component is transmitted along the surface of the earth and its presence is highly undesirable. An interesting property of the surface wave is that its velocity is significantly lower than that of the reflected wave. Velocity filtering can be applied to attenuate these undesirable signals.

Velocity filtering, here, will be performed by multidimensional filters included in the array processing scheme. Apparent velocity (see the velocity filtering example below, if unfamiliar with this term) of arriving signals depends upon the wave type, source location and structure beneath recording site. The typical signal and noise characteristics are summarized next.

(a) Teleseismic signals: These are assumed to propagate coherently across the array. Observed apparent velocities usually lie above 8 km/s (P-waves) and above 4 km/s (S-waves). P-waves show predominant frequencies around 1 Hz while S-waves usually have lower peak frequency.
(b) Microseismic noise: Propagate with velocities from about 2.5 to 4 km/s. Dominant frequencies occupy a broad low frequency range. Depending upon the

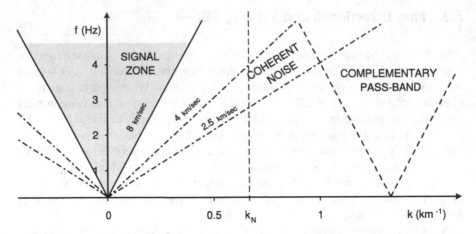

Fig. 5.3 Idealized distribution of the signal and noise components in the frequency vs wavenumber plane. Space sampling $\Delta x = 0.8$ km and time sampling $\Delta t = 0.1$ s are typical

distance to the source and sensor spacing, some portions of microseismic noise may propagate coherently across the array.

(c) Incoherent noise: Various human activities (factories, traffic, construction), action of wind, smaller water basins, etc., generate high frequency noise above 1 Hz. Corresponding apparent velocities vary for different sources, but in general have low values, around 1 km/s. For sensor spacing larger than several kilometers, it is most likely that the high frequency noise propagates incoherently across the array, irrespective of whether the location of the noise source is inside or outside the array pattern.

(d) Signal-generated noise: This is noise generated by the desired signal itself in the vicinity of the sensor as multiple reflections and mode conversions. It occupies a wide range of apparent velocities, depending upon the local structure and it may propagate coherently across the array.

In Fig. 5.3, P-waves with apparent velocities in the range from 8 km/s to infinity occupy the triangular zone labeled "signal zone." All kinds of coherent noise with apparent velocities between 2.5 and 4 km/s are distributed within the wedge marked "coherent noise."

5.3.1   Example of Implementation of Velocity Filtering

Consider the geometry shown in Fig. 5.4a. A line array of $(N + 1)$ equispaced detectors are positioned. A signal source is located at a distance d_0 from the first detector. Denoting by $s(t)$, the signal generated by the signal source, and ignoring

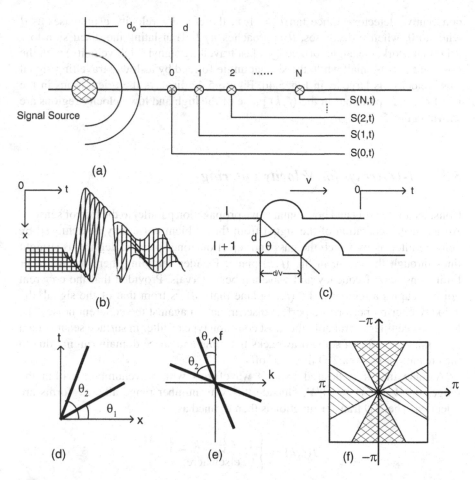

Fig. 5.4 The setting up of velocity filtering problem. (**a**) Signal and linear array of detectors; (**b**) typical signal trace; (**c**) outputs from the l and $l + 1^{th}$ sensor; (**d**) pattern in (x,t)-plane generated by two signals travelling with velocities $v_1, v_2, v_2 < v_1$; (**e**) pattern in (k–f) plane generated by the signals in (**d**); (**f**) velocity discrimination facilitated by fan-filtering

any attenuation by the medium of propagation, the output signal from the lth detector is denoted by

$$s(l, t) = s\left(t, \frac{d_0 + ld}{v}\right), l = 0, 1, \ldots, N$$

where v is the velocity of propagation. Figure 5.4b shows a typical 2-D trace generated by the linear array of detectors, where the temporal and spatial variables are denoted by t and x, respectively. The trace is composed of the output signals from the detectors. The output from the lth detector is the source signal delayed by $d_0 + ld/v, l = 0, 1, \ldots, N$. Figure 5.4c shows the output signals from two consecutive detectors. Since $\tan \theta = 1/v$, therefor the velocity v increases as θ tends to 0; when v decreases, θ approaches $\pi/2$. Translating the discussion to a 2-D framework, the angle formed by a fast travelling signal (high velocity) with the positive t-axis is small while the similar angle formed by a slowly travelling signal (low velocity) is large as in Fig. 5.4d. Figure 5.4e displays the orientations in the (f, k) plane. Separations in the (f, k) plane of the high and low velocity regions are shown in Fig. 5.4f.

5.3.2 1-D Arrays for Velocity Filtering

Consider a 1-D array and horizontal wave propagation parallel to the line of sensors. An adequate separation of the signal from the ambient noise may be achieved by velocity filtering by which the pass and rejection zones become defined as straight lines through the origin in the (f, k)-plane. Besides sampling there are no other limitations upon frequency and wavenumber intervals. Provided that the coherent noise occupies a region in the (f, k)-plane that differs from that of the signal, the velocity filtering performs a perfect discrimination against the coherent noise. The f–k pie slice filter is probably the most common type of filter in surface seismic data processing, in spite of some drawbacks like time and space domain ringing due to the creation of steps in 2-D filter cut-offs.

Assume that it is desired to pass waveforms with wavenumbers within the range, $-|f|/V \leq k \leq |f|/V$. Outside this wavenumber range all waveforms are rejection. The 2-D transfer function is then defined as:

$$H(f, k) = \begin{cases} 1, & -\frac{|f|}{V} \leq k \leq \frac{|f|}{V} \\ 0, & \text{elsewhere.} \end{cases} \tag{5.9}$$

The time-space impulse response of the filter may be expressed in terms of the inverse 2-D Fourier transform of H(f,k).

$$h(t, x) = \int_{-\infty}^{\infty} \int_{-\infty}^{\infty} H(f, k) \exp(j 2\pi (ft - kx)) df dk. \tag{5.10}$$

Due to the periodicity of $H(f, k)$ in wavenumber, we must limit ourselves to the resolvable wavenumber band only:

$$- k_N \leq k \leq k_N. \tag{5.11}$$

It follows from the basic relation,

$$f = Vk \tag{5.12}$$

that for a given apparent velocity V and frequencies $|f| > V k_N$ in the (f, k)-plane, we leave the primary and enter the complementary pass zone. Therefore, the frequency limits for a meaningful transfer function are:

$$-f_N \leq f \leq f_N, \tag{5.13}$$

where

$$f_N = V k_N = \frac{V}{2\Delta x}. \tag{5.14}$$

Having established the value of the folding frequency f_N, the sampling period Δ which preserves all information within the frequency band $(-f_N, f_N)$ becomes:

$$\Delta t = \frac{1}{2 f_N}. \tag{5.15}$$

The last equation also shows that the apparent out-off velocity satisfies:

$$V = \frac{f_N}{k_N} = \frac{\Delta x}{\Delta t}. \tag{5.16}$$

For any signal to pass through this velocity filter, the apparent velocity must be equal to or higher than V, i.e. for a given sensor spacing the signal move-out τ must satisfy the condition,

$$\tau \leq \Delta t. \tag{5.17}$$

Introducing the finite limits $\pm f_N$, $\pm k_N$, we have

$$h(t, x) = \int_{-f_N}^{f_N} \int_{-k_N}^{k_N} H(f, k) \exp(\jmath 2\pi (ft - kx)) df dk. \tag{5.18}$$

It may be expected intuitively that the sharpness between the pass and rejection zones increases and the amplitude of side lobes decreases with an increasing number of array sensors.

5.3.3 Velocity Filtering Example

The linear array of sensors in Fig. 5.5a is used for recording a shock wave, $s(t, x)$ where t is time and x is the distance measured from the center of the array. The shock wave propagates with an apparent (move-out) velocity, $V = v / \sin \phi$, on the earth's surface along the array line, where it is assumed that the plane wavefronts, travelling across the array sites at an angle ϕ with the horizontal, approach with a velocity v,

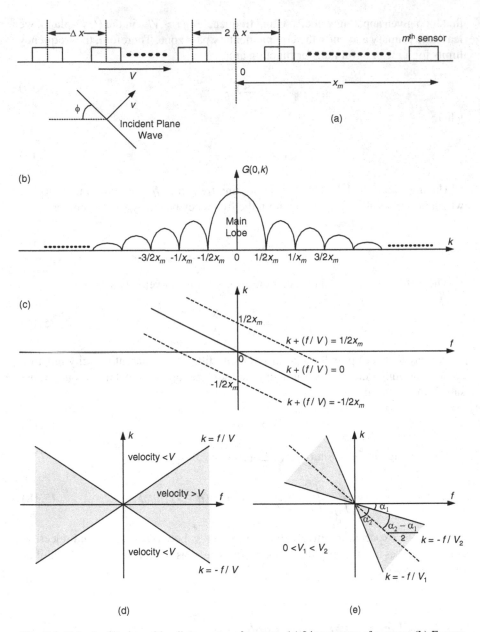

Fig. 5.5 Velocity filtering with a linear array of sensors. (**a**) Linear array of sensors. (**b**) Energy distribution (at zero temporal frequency) v/s spatial frequency. (**c**) Main energy distribution lines of received shock wave in (f, k)-plane. (**d**) Highpass velocity filter. (**e**) Bandpass velocity filter

and successively excite individual array sensors. The seismic signal, therefore, may be describable by

$$s(t,x) = \begin{cases} s\left(t - \frac{x}{V}\right), & |x| \leq x_m \\ 0, & |x| > x_m \end{cases}$$

Clearly, $s(t - (x/V))$ is a shifted version of $s(t) \triangleq s(t,0)$, the seismic record at the spatial origin; the shift equals x/V and represents the delay between the wave recorded at the center of the array and that recorded at a distance x from the center. The Fourier transform

$$H(f,k) = \int_{-\infty}^{\infty} \int_{-\infty}^{\infty} s(t,x) e^{-j2\pi(ft+kx)} dt\, dx$$

(it is assumed that the shock wave is measured continuously in $[-x_m, x_m]$) is the 2-D spectrum of $s(t,x)$ and f is the ordinary "temporal" frequency in Hertz, while the "spatial" frequency k is also referred to as the wavenumber. Let $y = t - (x/V)$ so that $t = y + (x/V)$.

$$H(f,k) = \int_{-\infty}^{\infty} \int_{-\infty}^{\infty} s(y) e^{-j2\pi\left[f\left(y+\frac{x}{V}\right)+kx\right]} dy\, dx$$

$$= \int_{-\infty}^{\infty} s(y) e^{-j2\pi fy} dy \int_{-x_m}^{x_m} e^{-j2\pi\left(\frac{f}{V}+k\right)x} dx$$

$$= S(f) \frac{\sin\left[2\pi\left(\frac{f}{V}+k\right)x_m\right]}{\pi\left(\frac{f}{V}+k\right)}$$

Let $f(t_1, t_2) \longleftrightarrow F(\omega_1, \omega_2)$ be a 2-D Fourier transform pair. Then the counterpart of Parseval's theorem for the 2-D case is:

$$E = \int_{-\infty}^{\infty} \int_{-\infty}^{\infty} |f(t_1, t_2)|^2 dt_1 dt_2 = \frac{1}{4\pi^2} \int_{-\infty}^{\infty} \int_{-\infty}^{\infty} |F(\omega_1, \omega_2)|^2 d\omega_1 d\omega_2$$

where E is the total signal energy. The spectrum, $F(\omega_1, \omega_2)$ gives the energy distribution. In this case, define

$$|H(f,k)|^2 \triangleq G(f,k).$$

For any fixed f, $G(f,k)$ is considered. For example, when $f = 0$, $G(0,k)$ has a typical shape given in Fig. 5.5b. Clearly, the main energy distribution is concentrated around $(f/V) + k = 0$ on the (f,k)-plane and the width of the band between the first two maxima is $1/x_m$, as is apparent from Fig. 5.5c. By shifting $G(0,k)$ by integer multiples of f/V along the frequency axis we can obtain a family of surfaces.

Assume that it is desired to pass only waveforms with wavenumbers within the range, $-(|f|/V) < k < (|f|/V)$. The transfer function of the so-called fan filter (pie-slice filter) required for the purpose is

$$H(f,k) = \begin{cases} 1, & -\frac{|f|}{V} < k < \frac{|f|}{V} \\ 0, & \text{otherwise.} \end{cases}$$

$H(f,k)$ defines a high-pass velocity filter which passes signals with apparent velocities of magnitude greater than V and rejects signals with lower velocities.

The shaded region in Fig. 5.5d, defined by $-(|f|/V) < k < (|f|/V)$, is obtained by taking the intersection of regions, $-(|f|/V) < k$ and $(|f|/V) > k$.

Consider a velocity pass range from V_1 to V_2 with a center velocity V_c. The filter which meets these specifications is a band-pass velocity filter. Rotation in the (f,k)-plane followed by a high-pass filtration may be used to replace band-pass filtering. The band-pass filter shown in Fig. 5.5e can be converted back to a high-pass filter by rotating anti-clockwise by $\frac{1}{2}(\alpha_1 + \alpha_2)$.

The result so far is based on the assumption of continuous spatial measurement. If that assumption is not made, it can be shown that

$$H(f,k) = 2S(f) \sum_{p=1}^{m} \cos\left[2\pi \left(k + \frac{f}{V}\right) p\Delta x\right]$$

How does the above expression change if there is an additional sensor located at the array center? That is, consider the case when there is an odd number, $2m + 1$, of sensors.

5.3.4 Impulse Response of Fan Filter

$$h[n_1, n_2] = \begin{cases} f_1[n_1, n_2], & n_1 = 0, n_2 = 0 \\ f_2[n_1, n_2], & n_1 \neq 0, n_2 = 0 \\ f_3[n_1, n_2], & n_1 + an_2 \neq 0, n_1 + bn_2 = 0 \\ f_4[n_1, n_2], & n_1 + an_2 = 0, n_1 + bn_2 \neq 0 \\ f_5[n_1, n_2], & \text{otherwise.} \end{cases}$$

Where,

$$f_1[n_1, n_2] = \frac{(a-b)B^2}{4\pi^2}$$

$$f_2[n_1, n_2] = \frac{(a-b)B^2}{2\pi^2 n_1^2}\left(n_1 B \sin(n_1 B) + \cos(n_1 B) - 1\right)$$

$$f_3[n_1, n_2] = \frac{1}{2\pi^2 n_2}\left(\frac{1 - \cos\left[(n_1 + an_2)B\right]}{n_1 + an_2}\right)$$

$$f_4[n_1, n_2] = \frac{1}{2\pi^2 n_2}\left(\frac{\cos\left[(n_1 + bn_2)B\right] - 1}{n_1 + bn_2}\right)$$

$$f_5[n_1, n_2] = \frac{1}{2\pi^2 n_2}\left(\frac{1 - \cos\left[(n_1 + an_2)B\right]}{n_1 + an_2} + \frac{\cos\left[(n_1 + bn_2)B\right) - 1}{n_1 + bn_2}\right].$$

Note that it is not possible to have both $n_1 + an_2 = 0$ and $n_1 + bn_2 = 0$, since this would imply that $a = b$, which cannot be true since it is given that $-1 < a < b < 1$. $a = b$ corresponds to the trivial case where the Fourier transform reduces to zero for all ω_1, ω_2, which would also mean that $h[n_1, n_2] = 0$ for all n_1, n_2.

5.3.5   Summary of 1-D Case

Parametric method of AR modelling is linked to linear prediction. Consider the 1-D case of a real discrete-time zero-mean WSS random process $x[n]$. By orthogonality principle (applied to the least mean-square method),

$$E\left\{\left[x[n] - \sum_{i=1}^{N} h[N; i]x[n - i]\right]x[n - k]\right\}$$

$$= \left[r[k] - \sum_{i=1}^{N} h[N; i]r[k - i]\right] = 0, \quad 1 \leqslant i \leqslant N \qquad (5.19)$$

The optimum mean-square prediction error is given by P_N below:

$$P_N = E\left\{\left[x[n] - \sum_{i=1}^{N} h[N; i]x[n - i]\right]^2\right\}$$

$$= \left[r[0] - \sum_{i=1}^{N} h[N; i]r[-i]\right] \qquad (5.20)$$

The preceding set of equations in (5.19) coupled with (5.20) lead to the *Yule-Walker* system ($r[i] = r[-i]$)

$$
\begin{bmatrix}
r[0] & r[1] & \cdots & r[N] \\
r[1] & r[0] & \cdots & r[N-1] \\
\vdots & & & \\
r[N] & r[N-1] & \cdots & r[0]
\end{bmatrix}
\begin{bmatrix}
1 \\
-h[N;1] \\
\vdots \\
-h[N;N]
\end{bmatrix}
=
\begin{bmatrix}
P_N \\
0 \\
\vdots \\
0
\end{bmatrix}
\tag{5.21}
$$

The $(N+1)$ equations can be solved for the $(N+1)$ unknowns, $h[N;1], \ldots, h[N;N], P_N$. Levinson's algorithm is an efficient way to implement the solution and the Levinson update scheme can be used to calculate the parameters of higher order predictors. A classical result in 1-D that does not fully generalize to 2-D is stated next.

Fact 5.2. *Assume that the covariance matrix in the preceding Toeplitz system of equations is positive definite. Then,*

(a) *There is a unique solution for PEF coefficients $\{h[N;1], \ldots, h[N;N]\}$ and PEF variance P_N*
(b) *$P_N > 0$*
(c) *The PEF*

$$
H_N(z) = \left[1 - \sum_{i=1}^{N} h[N;i]z^{-i} \right]
$$

is minimum phase i.e. the magnitudes of its poles and zeros are less than one.

5.3.6   2-D Counterpart

The notation is simplified to avoid clutter because you are now familiar with it, i.e. $h[k_1, k_2]$ is to be interpreted as $h[N_1; k_1, N_2; k_2]$. Then,

$$
x[n_1, n_2] = \sum_{(k_1,k_2)\in A} \sum h[k_1, k_2]x[n_1 - k_1, n_2 - k_2] + w[n_1, n_2]
$$

where $w[n_1, n_2]$ is a white noise process with variance σ^2. Suppose that the mask A is such that

$$
\text{supp}\{h[k_1, k_2]\} = \{0 \leqslant k_1 < N_1 \text{ and } 0 \leqslant k_2 < N_2\} \setminus \{0, 0\}
$$

Multiply both sides by $x^*[n_1 - l_1, n_2 - l_2]$ and apply the expectation operator to get

$$R_x[l_1, l_2] = \sum \sum h[k_1, k_2] R_x[l_1 - k_1, l_2 - k_2]$$
$$+ E\{w[n_1, n_2] x^*[n_1 - l_1, n_2 - l_2]\}, \qquad (5.22)$$

where $R_x[l_1, l_2]$ is the autocorrelation sequence. Note,

$$E\{w[n_1, n_2] x^*[n_1 - l_1, n_2 - l_2]\}$$
$$= \sum_{i_1} \sum_{i_2} h_A^*[i_1, i_2] E\{w[n_1, n_2] w^*[n_1 - l_1 - i_1, n_2 - l_2 - i_2]\}$$

where $\{h_A[i_1, i_2]\}$ is the impulse response of the IIR filter relating, ideally, $w[n_1, n_2]$ to $x[n_1, n_2]$ by convolution. Since $w[n_1, n_2]$ is a white noise process, therefore,

$$E\{w[n_1, n_2] w^*[n_1 - l_1, n_2 - l_2]\} = \sigma^2 \delta[l_1, l_2]$$

$$E\{w[n_1, n_2] x^*[n_1 - l_1, n_2 - l_2]\} = \sum_{i_1=-\infty}^{\infty} \sum_{i_2=-\infty}^{\infty} h_A^*[i_1, i_2] \delta[l_1 + i_1, l_2 + i_2]$$

$$= \sigma^2 h_A^*[-l_1, -l_2]$$

Therefore, Eq. (5.22) can be written as,

$$R_x[l_1, l_2] = \sum_{(k_1, k_2) \in A} \sum h[k_1, k_2] R_x[l_1 - k_1, l_2 - k_2] + \sigma^2 h_A^*[-l_1, -l_2]. \qquad (5.23)$$

Since the IIR filter has FQQP mask for denominator coefficients, a FQQP impulse response $\{h_A[i_1, i_2]\}$ is assumed. Restricting attention to the case when,

$$0 \leqslant l_1 < N_1 \text{ and } 0 \leqslant l_2 < N_2 = \text{supp}\{h[k_1, k_2]\} \cup \{0, 0\}$$

we have (since $h[0, 0] = 1$) from (5.23),

$$R_x[l_1, l_2] = \sum_{\substack{k_1=0 \\ (k_1, k_2) \neq (0,0)}}^{\infty} \sum_{k_2=0}^{\infty} h[k_1, k_2] R_x[l_1 - k_1, l_2 - k_2] + \sigma^2 \delta[l_1, l_2]$$

The above system may be written in matrix form

$$Ra = b \qquad (5.24)$$

where

$$\mathbf{a} = [1 - h[1,0] \cdots - h[N_1 - 1, 0] - h[0, 1] \cdots - h[N_1 - 1, N_2 - 1]]^T_{N_1 N_2 \times 1}$$

(after row-by-row scan starting from bottom row from left-to-right),

$$\mathbf{b} = [\sigma^2 0 \cdots 0 \cdots \cdots 0]^T_{N_1 N_2 \times 1},$$

R is Hermitian of order $N_1 N_2 \times N_1 N_2$.

If R is non-singular, then,

$$\mathbf{a} = R^{-1} \mathbf{b}.$$

The above system of equations in (5.24) also results from a linear prediction formulation.

$$\hat{x}[n_1, n_2] = \sum \sum_{(k_1, k_2) \in A} h[k_1, k_2] x[n_1 - k_1, n_2 - k_2]$$

where the prediction coefficients result from minimization of the error variance

$$\sigma^2 = E\left\{|x[n_1, n_2] - \hat{x}[n_1, n_2]|^2\right\}$$

Clearly,

Number of parameters to be determined $= N_1 N_2$
Number of known autocorrelation points $= (2N_1 - 1)(2N_2 - 1)$
Number of independent autocorrelation points $= 2N_1 N_2 - N_1 - N_2 + 1$

If the covariance matrix is positive definite, then there is a unique solution for the filter coefficients and σ^2 is always positive, as it must be. But the filter is not always minimum phase. Moreover, even if the filter is minimum phase, it can be shown that the transformation between the covariance matrix and the parameters is not invertible. Specifically, an infinite number of different positive definite covariance matrices can generate the same parameters.

5.4   Design of Multidimensional IIR Filters

5.4.1   A Spectral Transformation Method for Design of Multidimensional IIR Filters

The rectangular wavenumber response,

$$H(e^{j\omega_1}, e^{j\omega_2}) \triangleq H(\omega_1, \omega_2) = \begin{cases} 1, & |\omega_i| \leq \omega_{ci} < \pi \text{ for } i = 1, 2 \\ 0, & \text{otherwise} \end{cases}$$

has a separable unit impulse response

$$h[n_1, n_2] = \frac{1}{4\pi^2} \int_{-\omega_{c1}}^{\omega_{c1}} \int_{-\omega_{c2}}^{\omega_{c2}} e^{j\omega_1 n_1} e^{j\omega_2 n_2} \, d\omega_2 \, d\omega_1$$

$$= \left(\frac{\sin(\omega_{c1} n_1)}{\pi n_1} \right) \left(\frac{\sin(\omega_{c2} n_2)}{\pi n_2} \right)$$

which is of infinite extent. The unit impulse response of the filter with circularly symmetric wavenumber response

$$H(e^{j\omega_1}, e^{j\omega_2}) \triangleq H(\omega_1, \omega_2) = \begin{cases} 1, & \omega_1^2 + \omega_2^2 \leq \omega_c^2 < \pi^2 \\ 0, & \text{otherwise} \end{cases}$$

is also of infinite extent but non-separable. In particular,

$$h[n_1, n_2] = \frac{1}{4\pi^2} \int \int_{(\omega_1, \omega_2) \in [\omega_1^2 + \omega_2^2 \leq \omega_c^2]} e^{j(\omega_1 n_1 + \omega_2 n_2)} \, d\omega_2 \, d\omega_1$$

is convenient to evaluate after using the transformation,

$$\omega_1 = \omega \cos \theta, \qquad \omega_2 = \omega \sin \theta.$$

The determinant of the Jacobian matrix

$$J = \begin{bmatrix} \frac{\partial \omega_1}{\partial \omega} & \frac{\partial \omega_1}{\partial \theta} \\ \frac{\partial \omega_2}{\partial \omega} & \frac{\partial \omega_2}{\partial \theta} \end{bmatrix} = \begin{bmatrix} \cos \theta & -\omega \sin \theta \\ \sin \theta & \omega \cos \theta \end{bmatrix}$$

is $\det J = \omega$. Therefore, the expression for $h[n_1, n_2]$ can be rewritten in polar form as

$$h[n_1, n_2] = \frac{1}{4\pi^2} \int_0^{\omega_c} \int_0^{2\pi} \omega \, e^{j[\omega(n_1 \cos \theta + n_2 \sin \theta)]} \, d\omega \, d\theta$$

Let $n_1 = n \cos \phi$, $n_2 = n \sin \phi$ so that $\phi = \tan^{-1} \frac{n_2}{n_1}$ to get

$$h[n_1, n_2] = \frac{1}{4\pi^2} \int_0^{\omega_c} \int_0^{2\pi} e^{j\omega n \cos(\theta - \phi)} \, d\theta \, d\omega$$

$$= \frac{1}{2\pi} \int_0^{2\pi} \omega \, J_0(\omega \sqrt{n_1^2 + n_2^2}) \, d\omega$$

where $J_m(x)$ is the Bessel function (after Friedrich Wilhelm Bessel (1784–1846)) of the first kind and m-th order,

$$J_m(x) = \frac{1}{\pi} \int_0^{\pi} \cos(m\theta - x \sin \theta) \, d\theta$$

so that

$$
J_0(x) = \frac{1}{\pi} \int_0^\pi \cos(x \sin \theta) \, d\theta
$$

$$
= \frac{1}{2\pi} \int_0^{2\pi} \cos(x \sin \theta) \, d\theta = \frac{1}{2\pi} \int_0^{2\pi} \cos(x \cos \theta) \, d\theta
$$

Using the fact that

$$
x \, J_1(x) \, |_{x=a}^b = \int_a^b x \, J_0(x) \, dx
$$

$$
h[n_1, n_2] = \frac{\omega_c}{2\pi} \frac{J_1(\omega \sqrt{n_1^2 + n_2^2})}{\sqrt{n_1^2 + n_2^2}}
$$

Circularly symmetric filter response was approximated by the transformation technique for FIR filters in the previous chapter. In this chapter, the frequency transformation technique for design of a 2-D IIR filters uses the result quoted next. First, the characterization of a 2-D all pass function is given.

Definition 5.1. A real coefficient rational function expressible in reduced form as

$$
F(z_1, z_2) = \prod_{j=1}^d \left(z_1^{n_{1j}} z_2^{n_{2j}} \frac{\sum_{k_1=0}^{n_{1j}} \sum_{k_2=0}^{n_{2j}} a_j[k_1, k_2] z_1^{-k_1} z_2^{-k_2}}{\sum_{k_1=0}^{n_{1j}} \sum_{k_2=0}^{n_{2j}} a_j[k_1, k_2] z_1^{k_1} z_2^{k_2}} \right)
$$

is a bivariate all pass function. If the denominator of $F(z_1, z_2)$ is devoid of zeros outside the closed unit bidisc $\overline{U}^2$, when z_1^{-1} and z_2^{-1} are the delay variables, then $F(z_1, z_2)$ is a stable (in the BIBO sense, and, in this case, also structurally) all pass function. When $d = 1$, a basic all pass block is realized and when $d > 1$, this basic block is cascaded d times.

Fact 5.3. *Let $H(z_1, z_2)$ be the transfer function of a stable 2-D FQQP digital filter. If $F_1(z_1, z_2)$ and $F_2(z_1, z_2)$ are stable 2-D all pass functions, then, the transformations of the unit delay variables according to*

$$
z_1^{-1} \longrightarrow F_1(z_1, z_2), \qquad z_2^{-1} \longrightarrow F_2(z_1, z_2)
$$

in $H(z_1, z_2)$ result in a filter $H(F_1^{-1}(z_1, z_2), F_2^{-1}(z_1, z_2))$ which is stable.

Fig. 5.6 Portions of
wavenumber responses of
(from inside out) aster,
diamond, circularly
symmetric and square
wavenumber responses

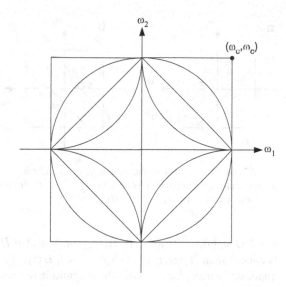

5.4.1.1 Diamond Shaped Filter

In the method of spectral transformation, a 2-D IIR filter is designed from a 1-D
IIR filter. Let $H(z)$ represent a 1-D causal and stable IIR filter. Two-dimensional
filters with diamond shaped wavenumber responses are known to preserve better,
the horizontal and vertical details, in comparison to circularly symmetric and square
separable filters having equivalent maximum cut-off frequency (ω_c, ω_c). The human
visual system is more sensitive to horizontal and vertical orientations than diagonal
ones. The diamond-shaped wavenumber response can be considered as a special
case of aster shaped response as illustrated in Fig. 5.6.

The 1-D causal prototype filter, say, is a low pass filter with cutoff at $\pi/2$
radians/s. We design a 2-D filter

$$H_1(z_1, z_2) = H(z_1)H(z_2)$$

$$H_2(z_1, z_2) = H(z_1 z_2)H(z_1 z_2^{-1})$$

Then, $H_1(z_1, z_2)$ must be stable and FQQP recursible. The magnitude of wavenumber response of $H_2(z_1, z_2)$ approximates a diamond shaped wavenumber response.
For plotting purposes, the ideal responses are used. Note that $z_1 z_2$ and $z_1 z_2^{-1}$ are
all pass functions. Therefore, if the delay variables in $H(z_1, z_2)$ are transformed
according to

$$z_1^{-1} \longrightarrow z_1^{-1} z_2^{-1}, \qquad z_2^{-1} \longrightarrow z_1^{-1} z_2$$

then, $H_2(z_1, z_2)$ results. Since $z_1^{-1} z_2^{-1}$ and $z_1^{-1} z_2$ are stable all pass functions with no
non-essential singularities of the second kind on T^2, 1-D stability of $H(z)$ implies

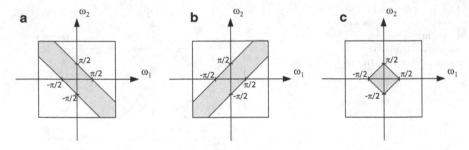

Fig. 5.7 (**a**) The support of the wavenumber magnitude response of $H(z_1 z_2)$; (**b**) the support of the wavenumber magnitude response of $H(z_1^{-1} z_2)$ and (**c**) the realized diamond shaped response obtained by cascading $H(z_1 z_2)$ and $H(z_1^{-1} z_2)$

the 2-D stability of not only $H_1(z_1, z_2)$, but also $H_2(z_1, z_2)$. Furthermore, if $H(z)$ is causal, then $H_1(z_1, z_2)$ is FQQP causal, $H(z_1 z_2)$ is FQQP and $H(z_1^{-1} z_2)$ is fourth quadrant quarter plane causal. The magnitude responses of $H(z_1 z_2)$ and $H(z_1^{-1} z_2)$ are sketched in Fig. 5.7a, b. The support of these responses are identified by the shaded area. It is then clear that when $H(z_1 z_2)$ and $H(z_1^{-1} z_2)$ are cascaded, the resulting transfer function $H_2(z_1, z_2)$ realizes a diamond shaped pass band.

We design another 2-D filter $H_T(z_1, z_2)$ by

$$H_t(z_1, z_2) = H_1(z_1, z_2) H_2(z_1, z_2)$$

What is the approximate magnitude of the wavenumber response of $H_T(z_1, z_2)$?

5.4.2   Whitening Filter and Wiener Filter

The support of the unit impulse response sequence of a 2-D asymmetric half-plane filter can be ordered as follows. For any two pair $[m_1, m_2]$, its past is the set of points

$$\{[k_1, k_2]   :   k_1 = m_1, \ k_2 < m_2; \ k_1 < m_1, \ -\infty < k_2 < \infty\}$$

and its future is the set of points

$$\{[k_1, k_2]   :   k_1 = m_1, \ k_2 > m_2; \ k_1 > m_1, \ -\infty < k_2 < \infty\}$$

After totally ordering the points in the support of the impulse response, if $[k_1, k_2]$ is in the past of $[m_1, m_2]$ use the notation

$$[k_1, k_2] < [m_1, m_2] \ \text{ or } \ [m_1, m_2] > [k_1, k_2].$$

Consider the whitening filter process,

$$w[m_1, m_2] = \sum_{(k_1, k_2) \geq (0,0)} \sum h[k_1, k_2] x[m_1 - k_1, m_2 - k_2] \qquad (5.25)$$

$$E\{w[m_1 + k_1, m_2 + k_2][w[m_1, m_2]\} = \sigma^2 \delta[k_1, k_2], \qquad (5.26)$$

where $h[k_1, k_2], x[m_1, m_2]$ are, respectively, the unit impulse response of the linear shift-invariant whitening filter and an input zero-mean real-valued discrete random process. Furthermore, $w[m_1, m_2]$ is the desired zero-mean real-valued white-noise process, with autocorrelation function, $\sigma^2 \delta[k_1, k_2]$.

Substitute the first equation into the second and move the expectation operator inside the summations (assuming this is possible to do in case of infinite sums) so as to operate on the random variables setting up the random field to get ($h[k_1, k_2]$ is **not** a random variable)

$$\sigma^2 \delta[m_1, m_2] = \sum_{(k_1, k_2) \geq (0,0)} \sum \sum_{(l_1, l_2) \geq (0,0)} \sum h[k_1, k_2] h[l_1, l_2] r[m_1 - k_1 + l_1, m_2 - k_2 + l_2],$$
$$(5.27)$$

where

$$r[k_1, k_2] = E\{x[m_1 + k_1, m_2 + k_2] x[m_1, m_2]\} \qquad (5.28)$$

Taking the z-transform of both sides of (5.27) after multiplying left and right hand sides by $z_1^{-m_1} z_2^{-m_2}$ and summing over $[-\infty, \infty] \times [-\infty, \infty]$ i.e. the $\mathbb{Z} \times \mathbb{Z}$ index set, one gets

$$\sigma^2 = S(z_1, z_2) H(z_1, z_2) H(z_1^{-1}, z_2^{-1}), \qquad (5.29)$$

where $S(z_1, z_2) \leftrightarrow r[k_1, k_2]$ are z-transform pairs.

To solve the preceding equation for $H(z_1, z_2)$, it is required to consider the *spectral factorization problem*. In the univariate counterpart, an unique minimum-phase solution of finite order can be constructed for a rational power spectral density that is positive-valued on the unit circle. In the bivariate case, significant differences occur because of the fact that a minimum-phase spectral factor $H(z_1, z_2)$ of finite order does not exist, in general.

5.4.2.1 Properties of Whitening Filter $H(z_1, z_2)$

A. $H(z_1, z_2)$ is unique to within a multiplicative constant.

Proof. Suppose it is possible to have two solutions given by

$$S(z_1, z_2) = \frac{\sigma^2}{H(z_1, z_2) H(z_1^{-1}, z_2^{-1})} = \frac{\hat{\sigma}^2}{\hat{H}(z_1, z_2) \hat{H}(z_1^{-1}, z_2^{-1})}. \qquad (5.30)$$

Then,

$$\frac{\hat{H}(z_1, z_2)\sigma^2}{H(z_1, z_2)} = \frac{H(z_1^{-1}, z_2^{-1})\hat{\sigma}^2}{\hat{H}(z_1^{-1}, z_2^{-1}).} \tag{5.31}$$

The left-hand side of above is a weakly causal filter while the right-hand side is a weakly anticausal filter. Clearly, this is possible only if each side of the equation is a constant. Therefore,

$$\hat{H}(z_1, z_2) = cH(z_1, z_2) \tag{5.32}$$

where c is a constant.

B. $H(z_1, z_2)$ is proportional to the least-squares linear prediction error filter for $x[k_1, k_2]$ given the infinite past.

Proof. Normalize the whitening filter as follows.

$$H_N(z_1, z_2) = \frac{1}{h[0, 0]}H(z_1, z_2)$$

$$= [1 - \sum_{(k_1, k_2) > (0,0)} \sum h_N[k_1, k_2]z_1^{-k_1} z_2^{-k_2}] \tag{5.33}$$

where

$$S(z_1, z_2) = \frac{\sigma_N^2}{H_N(z_1, z_2)H_N(z_1^{-1}, z_2^{-1})} \text{ and } \sigma_N^2 = \frac{\sigma^2}{h^2[0, 0].} \tag{5.34}$$

Cross-multiplying,

$$H_N(z_1, z_2)S(z_1, z_2) = \frac{\sigma_N^2}{H_N(z_1^{-1}, z_2^{-1})}. \tag{5.35}$$

Taking the inverse z-transform and using the fact that the right-hand side is a weakly anticausal filter, we have

$$r[m_1, m_2] - \sum_{(k_1, k_2) > (0,0)} \sum h_N[k_1, k_2]r[m_1 - k_1, m_2 - k_2] = \sigma_N^2\delta[m_1, m_2] \tag{5.36}$$

$$(m_1, m_2) \geq (0, 0)$$

or

$$E\{[x[k_1, k_2] - \sum_{(l_1, l_2) > (0,0)} \sum h_N[l_1, l_2]x[k_1 - l_1, k_2 - l_2]]x[k_1 - m_1, k_2 - m_2]\}$$

$$= \sigma_N^2\delta[m_1, m_2], (m_1, m_2) \geq (0, 0). \tag{5.37}$$

The above equation satisfies **the orthogonality principle** which says that $H_N(z_1, z_2)$ operates on the random process $x[k_1, k_2]$ to produce a white-noise process that is uncorrelated with all past values of $x[k_1, k_2]$.

5.4.2.2 The 2-D Wiener Filtering Problem

We observe a signal $x[k_1, k_2]$, which is the sum of a message, $s[k_1, k_2]$ and noise $n[k_1, k_2]$. We model the message and noise as a wide-sense stationary random process and we assume that the power density spectra are known. The problem is to design a filter $G(z_1, z_2)$ which will give the optimum linear least-squares estimate for the message. 2-D Wiener filtering is of interest in image processing as well as array processing.

The classical solution to the problem involves first finding the minimum phase whitening filter for the observed signal by solving the corresponding spectral factorization problem. The idea is that the optimum filter $G(z_1, z_2)$ can be represented as the product of the whitening filter $H(z_1, z_2)$ and some other filter $H_a(z_1, z_2)$.

$H(z_1, z_2)$ operates on the random process $x[k_1, k_2]$ to produce a white-noise process that is uncorrelated with all past values of $x[k_1, k_2]$. Therefore, the orthogonality principle is satisfied; so $H(z_1, z_2)$ is the least squares prediction error filter for $x[k_1, k_2]$ given the infinite past. It was seen that

$$S(z_1, z_2) = \frac{\sigma^2}{H(z_1, z_2)H(z_1^{-1}, z_2^{-1})},$$

where $H(z_1, z_2)$ is the 2-D z-transform of $\{h[k_1, k_2]\}$.

5.4.2.3 Spectral Factorization and Hilbert Transform

The method for obtaining the spectral factorization via the Hilbert transform is applicable when the rational power spectrum density, analytic in some neighbourhood of T^2, obtained by evaluating $S(z_1, z_2)$ on T^2, is strictly positive on T^2. That is

$$S\left(e^{jw_1}, e^{jw_2}\right) > 0 \text{ for all } w_1, w_2$$

induces a factorization of the type

$$S(z_1, z_2) = G(z_1, z_2)G(z_1^{-1}, z_2^{-1}),$$

where $G(z_1, z_2)$ will be shown to be spectral factor. Given a positive "analytic" spectrum, it can be shown that the complex logarithm of the spectrum is analytic in some neighborhood of T^2. Therefore, $S(z_1, z_2)$ has a Laurent series expansion in that region:

$$\log S(z_1, z_2) = \sum_{k_1=-\infty}^{\infty} \sum_{k_2=-\infty}^{\infty} r[k_1, k_2] z_1^{-k_1} z_2^{-k_2}$$

where

$$r[k_1, k_2] = r[-k_1, -k_2] = \left(\frac{1}{2\pi j}\right)^2 \int \int_{T^2} z_1^{k_1-1} z_2^{k_2-1} \log S(z_1, z_2) dz_1 dz_2.$$

The decomposition,

$$\log S(z_1, z_2) = C(0, 0) + C(z_1, z_2) + C(\frac{1}{z_1}, \frac{1}{z_2}),$$

is possible, where

$$C(0, 0) = r[0, 0] > 0,$$

$$C(z_1, z_2) = \sum \sum_{(k_1, k_2) > (0,0)} r[k_1, k_2] z_1^{-k_1} z_2^{-k_2}.$$

Therefore, $S(z_1, z_2)$ can be expressed in the form

$$S(z_1, z_2) = e^{C(0,0)} A(z_1, z_2) A(z_1^{-1}, z_2^{-1})$$

where

$$A(z_1, z_2) = e^{C(z_1, z_2)}$$

$$= \sum_{k=0}^{\infty} \frac{1}{k!} [C(z_1, z_2)]^k$$

$$= \left[1 + \sum \sum_{(k_1, k_2) > (0,0)} r[k_1, k_2] z_1^{-k_1} z_2^{-k_2} \right].$$

Since $C(z_1, z_2)$ is analytic, it follows that $A(z_1, z_2)$ is analytic as well and from the series expansion it is weakly causal. $A(z_1, z_2)$ is analytic minimum phase because its inverse, $A^{-1}(z_1, z_2) = \exp[-C(z_1, z_2)]$ is also analytic and weakly causal. The spectral factor $G(z_1, z_2)$ is then identified to be a constant multiplier of $A(z_1, z_2)$.

Consider a 2-D PSD $S(z_1, z_2)$

$$S(z_1, z_2) = \sum_{k_1=-N}^{N} \sum_{k_2=-M}^{M} r[k_1, k_2] z_1^{-k_1} z_2^{-k_2}$$

$$= G(z_1, z_2) G(z_1^{-1}, z_2^{-1})$$

where

$$G(z_1, z_2) = \sum\sum_{(k_1,k_2)\geq(0,0)} g[k_1, k_2] z_1^{-k_1} z_2^{-k_2}$$

$$G(z_1, z_2) = G^{-1}(\frac{1}{z_1}, \frac{1}{z_2}) S(z_1, z_2),$$

and

$$G^{-1}(z_1, z_2) = \sum\sum_{(k_1,k_2)\geq(0,0)} \mu[k_1, k_2] z_1^{-k_1} z_2^{-k_2}.$$

Then, it follows that

$G(z_1, z_2)$

$$= \sum\sum_{(k_1,k_2)\leq(0,0)} \mu[-k_1, -k_2] z_1^{-k_1} z_2^{-k_2} \left[\sum_{k_1=-N}^{N} \sum_{k_2=-M}^{M} r[k_1, k_2] z_1^{-k_1} z_2^{-k_2} \right]$$

$$= \left[\sum\sum_{(k_1,k_2)\leq(0,0)} \mu[-k_1, -k_2] z_1^{-k_1} z_2^{-k_2} \right]$$

$$\left[\sum\sum_{(-N,-M)\leq(k_1,k_2)\leq(N,M)} r[k_1, k_2] z_1^{-k_1} z_2^{-k_2} \right].$$

Therefore, $G(z_1, z_2)$, in this case, takes the form,

$$G(z_1, z_2) = \sum\sum_{(0,0)\leq(k_1,k_2)\leq(N,M)} g[k_1, k_2] z_1^{-k_1} z_2^{-k_2}.$$

Note that you have to get $S(z_1, z_2)$ and σ^2 by *spectrum estimation*.

Example. Consider the power spectral density function $S(z_1, z_2)$, which is the z-transform of the autocorrelation sequence, $\{r[k_1, k_2]\}$.

$$S(z_1, z_2) = \sum\sum r[k_1, k_2] z_1^{-k_1} z_2^{-k_2}$$

$$= c + [z_2 + z_2^{-1}] + [z_1 + z_1^{-1}], \quad c > 4.$$

Clearly, $S(z_1, z_2) > 0$ on T^2. Consider,

$$S(z_1, e^{jw_2}) = (c + 2\cos w_2) + (z_1 + z_1^{-1}),$$

which for an arbitrary but fixed w_2 has a spectral factorization of the form,

$$S(z_1, e^{jw_2}) = (A(w_2)z_1 + B(w_2)) + (A(w_2)z_1^{-1} + B(w_2)),$$

$$A(w_2)B(w_2) = 1, \quad A^2(w_2) + B^2(w_2) = c + 2\cos w_2 > 0.$$

Therefore, combining the constraints in the last line,

$$A^4(w_2) - (c + 2\cos w_2)A^2(w_2) + 1 = 0,$$

which from the implicit function theorem must have a solution that satisfies,

$$2A^2(w_2) = c + 2\cos w_2 + \sqrt{(c + 2\cos w_2)^2 - 4},$$

where the right-hand side is guaranteed to be positive so that the solution $A(w_2)$ is a real-valued periodic function of w_2. By Fourier series expansion

$$A(w_2) = \sum_{k_2=-\infty}^{\infty} g_{k_2} e^{jk_2 w_2}, \quad g_{k_2} = g_{-k_2},$$

where the g_k's are the Fourier coefficients so that $S(z_1, z_2)$ has a spectral factor of infinite order in z_2 and finite order in z_1. This can be further justified by noting that the z-transform pair,

$$k_1 r[k_1, k_2] \leftrightarrow -z_1 \frac{\partial}{\partial z_1} S(z_1, z_2) = \frac{z_1^{-1} - z_1}{S(z_1, z_2)}$$

implies that

$$r[k_1, k_2] = r_1[k_1 - 1, k_2] - r_1[k_1 + 1, k_2]$$

where by taking 1-D z-transform at a time,

$$\sum_{k_2=-\infty}^{\infty} r_1[k_1, k_2] z_2^{-k_2} \triangleq \sum_{k_2=-\infty}^{\infty} \left(\sum_{k_1=-\infty}^{\infty} k_1 r[k_1, k_2] z_1^{-k_1} \right) z_2^{-k_2} = \frac{z_1^{-1} - z_1}{S(z_1, z_2)}.$$

Therefore, for an arbitrary but fixed integer k_1

$$r[k_1, k_2] = r[k_1, -k_2],$$

consistent with the constraint on the Fourier coefficients of $A(w_2)$.

Note that in the factor, $A(w_2) + B(w_2)z_1^{-1}$, since $A(w_2)$ is real valued, therefore the support of the sequence $\{h[k_1, k_2]\}$ in

$$S(z_1, z_2) = H(z_1, z_2)H(z_1^{-1}, z_2^{-1})$$

$$H(z_1, z_2) \triangleq \sum\sum h[k_1, k_2] z_1^{-k_1} z_2^{-k_2}$$

$$= \sum_{k_2=-\infty}^{\infty} g_{k_2} z_2^{-k_2} + \frac{1}{\displaystyle\sum_{k_2=-\infty}^{\infty} g_{k_2} z_2^{-k_2}} z_1^{-1},$$

is, necessarily,

$$supp\{h[k_1, k_2]\} = \{[k_1, k_2] : k_1 = 0, k_2 = 0, 1, \ldots, \infty;$$
$$k_1 = 1, k_2 = -\infty, \ldots, -1, 0, 1, \ldots, \infty\}.$$

Indeed, arguments can be completed to show that $H(z_1, z_2)$ is spectral factor.

5.4.2.4   The 2-D Complex Cepstrum

$$\hat{x}[k_1, k_2] = \frac{1}{(2\pi)^2} \int_0^{2\pi} \int_0^{2\pi} \ln X(w_1, w_2) e^{j(w_1 k_1 + w_2 k_2)} dw_1 \, dw_2.$$

$\hat{X}(w_1, w_2) = \ln X(w_1, w_2)$ must be continuous, differentiable, and doubly periodic and the function $\ln X(z_1, z_2)$ must be analytic in some region of convergence $\hat{R}$, where $R \supset \hat{R} \supset T^2$, with R the ROC of $X(z_1, z_2)$. Clearly, the complex cepstrum exists only if $X(w_1, w_2) \neq 0$ on the torus (distinguished boundary of the unit bidisc. Furthermore, a bisequence $x[k_1, k_2]$ that has a real and stable $\hat{x}[k_1, k_2]$ is said to have a *valid complex cepstrum*. The function,

$$\hat{X}(w_1, w_2) \triangleq \ln |X(w_1, w_2)| + j\phi(w_1, w_2),$$

where

$$X(w_1, w_2) = |X(w_1, w_2)| + e^{j\phi(w_1, w_2)}$$

is continuous and differentiable if $\phi(w_1, w_2)$ is the **unwrapped phase**. It can be shown that , in general, this unwrapped phase function is the sum of a doubly period phase function and a function linear in w_1 and w_2.

For the sake of simplicity, but without loss of generality, we assume that $x[k_1, k_2]$ is of finite support, so that

$$X(w_1, w_2) = \sum_{k_1} \sum_{k_2} x[k_1, k_2] e^{-j(w_1 k_1 + w_2 k_2)}$$

is a bivariate trigonometric polynomial. First, consider

$$A_{w_1}(z_2) = X(w_1, w_2)\Big|_{\substack{w_1 \text{ fixed} \\ e^{jw_2} \to z_2}} = \sum_{k_1} \sum_{k_2} x[k_1, k_2] z_2^{-k_2} e^{-jw_1 k_1}$$

as an univariate polynomial in z_2^{-1}, parametrized by w_1. Viewed as a rational function in z_2, $A_{w_1}(z_2)$ will have poles inside $|z_2| \leq 1$ only at $z_2 = 0$. This may be removed by multiplication with an appropriate power N_2 of z_2 to form the polynomial in z_2,

$$C_{w_1}(z_2) = z_2^{N_2} A_{w_1}(z_2).$$

Now, let us define the phase function as a contour integral, for an arbitrary but fixed w_1.

$$\phi(w_1, w_2) = Im \int_{z_2 = e^{j0} = 1}^{e^{jw_2}} \frac{A'_{w_1}(z_2)}{A_{w_1}(z_2)} dz_2 + \phi(w_1, 0)$$

where the univariate phase function $\phi(w_1, 0)$ is the phase as a function of w_1 for $w_2 = 0$. Letting,

$$X(w_1, 0) = X_R(w_1, 0) + j X_I(w_1, 0),$$

$$\phi(w_1, 0) = \int_0^{w_1} \left. \left(\frac{\frac{\partial X_I}{\partial w_1} X_R - \frac{\partial X_R}{\partial w_1} X_I}{X_R^2 + X_I^2} \right) \right|_{w_2=0} dw_1$$

Then, $\phi(w_1, w_2)$, is continuous and odd($\phi(w_1, w_2) = \phi(-w_1, -w_2)$), and we can write

$$\phi(w_1, 2\pi) = Im \left\{ \oint \frac{A'_{w_1}(z_2)}{A_{w_1}(z_2)} dz_2 \right\}$$

$$= Im \left\{ \oint \left(-N_2 z_2^{-1} + \frac{C'_{w_1}(z_2)}{C_{w_1}(z_2)} \right) dz_2 \right\}$$

$$= -2\pi N_2 + 2\pi r_2.$$

The last equation follows from application of residue theorem where r_2 is the number of roots (including multiplicities) of $C_{w_1}(z_2)$ and, therefore, of $A_{w_1}(z_2)$ inside the unit circle, $|z_2| \le 1$. If we let $k_{w_2} = r_2 - N_2$, then

$$\phi(w_1, w_2 + 2\pi) = \phi(w_1, w_2) + 2\pi k_{w_2}$$

Similarly, we can derive that

$$\phi(w_1 + 2\pi, w_2) = \phi(w_1, w_2) + 2\pi k_{w_1}$$

What remains to be shown is that k_{w_2} is not a function of w_1 and k_{w_1} is not a function of w_2. If we examine the roots of $C_{w_1}(z_2)$, which are also the roots of $A_{w_1}(z_2)$, as we continuously vary the parameter w_1 from 0 to 2π, we discover that the roots move in a continuous manner. Thus, for a root to move from inside to outside the unit circle (or vice versa), it must lie on the unit circle, $|z_2| = 1$, for some value of w_1. This, however, violates the hypothesis $X(w_1, w_2) \ne 0$. Therefore, given a continuous odd function $\phi(w_1, w_2)$ such that

$$\phi(w_1, w_2 + 2\pi) = \phi(w_1, w_2) + 2\pi k_{w_2}$$

$$\phi(w_1 + 2\pi, w_2) = \phi(w_1, w_2) + 2\pi k_{w_1}$$

$$\phi(w_1, w_2) = -\phi(-w_1, -w_2),$$

we can subtract the linear phase component

$$\phi_L(w_1, w_2) = k_{w_1} w_1 + k_{w_2} w_2$$

to give the remaining term

$$\phi_A(w_1, w_2) = \phi(w_1, w_2) - \phi_L(w_1, w_2)$$

Setting, $k_{w_2} \to K_2, k_{w_1} \to K_1$, the sequence

$$y[k_1, k_2] = x[k_1 - K_1, k_2 - K_2]$$

which is a shifted version of $x[k_1, k_2]$ has a Fourier transform,

$$Y(w_1, w_2) = X(w_1, w_2)e^{-j(w_1 K_1 + w_2 K_2)}.$$

Consequently, the unwrapped phase of $Y(w_1, w_2)$ is $\phi(w_1, w_2) - w_1 K_1 - w_2 K_2$, is continuous and doubly periodic, and $Y(w_1, w_2)$ satisfies the conditions necessary to define the complex cepstrum $\hat{y}[n_1, n_2]$.

Chapter 6
Wavelets and Filter Banks

6.1 Historical Results Leading up to Wavelets

Denote by $L_2[0, 2\pi]$, the space of classes of functions that are square-integrable on $[0, 2\pi]$. The sequence of functions $1/\sqrt{2\pi}, \cos x/\sqrt{\pi}, \sin x/\sqrt{\pi},$ $\cos 2x/\sqrt{2\pi}, \sin 2x/\sqrt{2\pi}, \ldots$ forms an orthonormal basis for $L_2[0, 2\pi]$.

6.1.1 Trigonometric Fourier Series (1807)

Let $f(x)$ be 2π-periodic. Then

$$f(x) = a_0 + \sum_{k=1}^{\infty}(a_k \cos kx + b_k \sin kx)$$

$$a_0 = \frac{1}{2\pi}\int_0^{2\pi} f(x)\, dx$$

$$a_k = \frac{1}{\pi}\int_0^{2\pi} f(x) \cos kx\, dx$$

$$b_k = \frac{1}{\pi}\int_0^{2\pi} f(x) \sin kx\, dx$$

Paul Du Bois-Raymond (1873) constructed a continuous 2π-periodic function of the real variable x, where its Fourier series diverged at a given point. Henri Lebesgue showed the space $L_2[0, 2\pi]$ of functions that are square-summable on the interval $L_2[0, 2\pi]$ has a Fourier series which converges to f in the sense of the quadratic mean.

© Springer International Publishing AG 2017
N.K. Bose, *Applied Multidimensional Systems Theory*,
DOI 10.1007/978-3-319-46825-9_6

Modification of definition of convergence by replacing partial sum $S_n(x)$ with Cesaro sums $\sigma_n = [S_0 + \cdots + S_n - 1]/n$ makes everything fall in place, overcoming the drawback pointed out by Du Bois-Raymond concerning Fourier's assertion that any 2π-periodic function converges pointwise to its Fourier series.

6.1.2 Exponential Fourier Series (1807)

Any function $f \in L_2[0, 2\pi]$ has expansion

$$f(x) = \sum_{n=-\infty}^{\infty} c_n e^{jnx},$$

where

$$c_n = \frac{1}{2\pi} \int_0^{2\pi} f(x) e^{-jnx} \, dx.$$

Convergence:

$$\lim_{M,N \to \infty} \int_0^{2\pi} |f(x) - \sum_{n=-M}^{N} c_n e^{jnx}|^2 \, dx = 0.$$

Fact 6.1. *Every 2π-periodic square integrable function is generated by a superposition of integral dilations $w_x(x) = w(nx)$ of the basis function $w_2(x) = e^{jx}$. From the orthonormal property of $\{w_n\}$, it follows that*

$$\frac{1}{2\pi} \int_0^{2\pi} |f(x)|^2 \, dx = \sum_{n=-\infty}^{\infty} |c_n|^2$$

Definition 6.1. Let $s(x) = \sum_{n=1}^{\infty} u_n(x)$ be an infinite sum of a sequence of functions that converge to $s(x)$ for each value of x in $[a, b]$. Denote the partial sum and pointwise remainder by

$$s_n(x) = u_1(x) + u_2(x) + \cdots + u_n(x)$$

$$r_n(x) = s(x) - s_n(x).$$

Then ordinary convergence is defined as: $\lim r_n(x) = 0$, because $\lim s_n(x) = s(x)$.

Definition 6.2. The series $\sum_{n=1}^{\infty} u_n(x)$ is uniformly convergent *in the interval $[a, b]$ if for each $\epsilon > 0$, there is a number N independent of x such that $|r_n(x)| < \epsilon$ for all $n > N$.*

The next question is: How do we represent a function $f(x)$ in space $L_2(\mathbb{R})$, i.e. $\int_{-\infty}^{\infty} |f(x)|^2\, dx < \infty$? The answer is to look for small waves or wavelets to generate $L_2(\mathbb{R})$. For $L_2(\mathbb{R})$, seek mother wavelet $\psi(x)$ with integral shifts $\psi(x-k), k \in \mathbb{Z}$, and dilation (for frequency bands) to give $\psi(2^j x - k), j, k \in \mathbb{Z}^+$.

6.1.3 Haar System (1909)

Does there exist another orthonormal system $h_0(x), h_1(x), \ldots, h_n(x), \ldots$ defined on $[0, 1]$, such that for any function $f(x)$ continuous on $[0, 1]$, the series

$$\sum_{k=0}^{\infty} \langle f, h_k \rangle h_k(x), \text{ where } \langle f, h_k \rangle = \int_0^1 f(x)\overline{h_k(x)}\, dx$$

converges uniformly to $f(x)$ on $[0, 1]$?

Haar's solution is given next.

$$h(x) = \begin{cases} 1, 0 \leq x < 1/2 \\ -1, 1/2 \leq x < 1 \\ 0, otherwise \end{cases}$$

For $n \geq 1$, write $n = 2^j + k$, $j \geq 0, 0 \leq k < 2^j$. Note that the index n ranges over all positive integers.

$$\psi_n(x) \triangleq 2^{j/2} h(2^j x - k) = h(j, k, x)$$

$$\text{Supp}\left\{ \psi_n(x) \triangleq h(j, k, x) \right\} \text{ is } I_n = [k2^{-j}, (k+1)2^{-j}]$$

To complete the set, define $\phi(x) \triangleq 1$ on $[0, 1)$.

Then, the sequence $\{\phi(x), \psi_1(x), \ldots, \psi_n(x)\}$ is an orthonormal basis for $L_2[0, 1]$ (Fig. 6.1).

6.1.4 Gabor STFT Transform

This is in contrast to Fourier transform, where the expansion is done by functions like $e^{-j\omega t}$ which are *not time-limited*, but highly localized in frequency. Recall that a Fourier series results from *integer dilations* of e^{jt} while a Fourier transform occurs from non-integer dilations of e^{jt}. The STFT of the function $p(t)$ involves multiplication by a sliding window $w(t - \beta)$ to localize the time domain data. The STFT then is

$$P_{STFT}(\omega, p) = \int_{-\infty}^{\infty} p(t)w(t - \beta)e^{-j\omega t} dt.$$

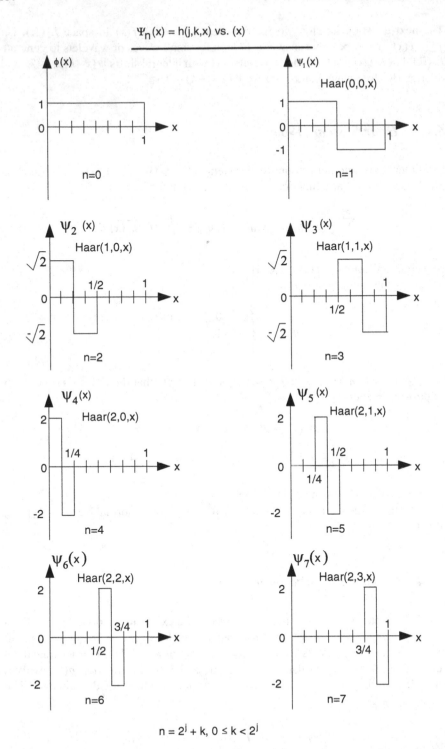

Fig. 6.1 Note that $\psi(2^j x - k)$ is obtained from $\psi(x)$ by a binary dilation of 2^j and a dyadic translation of $k/2^j$

The problem here is that localization in time by a narrow window, the localization in the frequency domain is compromised by the uncertainty principle. To obtain the original signal back, assume, without loss of generality, that $w(0) \neq 0$, so that after setting $\beta = t$

$$p(t)w(0) = \frac{1}{2\pi} \int_{-\infty}^{\infty} P_{STFT}(w, t)e^{j\omega t} d\omega$$

Observe that to recover $p(t)$ from P_{STFT}, it is not required to know $w(t)$ for all t. Gabor (1946) attempted to develop a methodology for localizing a function in time and frequency. This allows tracking the frequency content of a signal by observing the response in certain narrow frequency bands.

6.1.5   Continuous Gabor-Like Transforms (CGT)

CGT is described in terms of a representation of a signal with respect to all the *translations* and *modulations* of a *fixed-size window* function $q \in L_2(\mathbb{R}) \backslash \{0\}$. In fact, it is conventional to impose more constraints on q so that $q \in L_1 \cap L_2$. We can show that

$$q(t) \in L_2(\mathbb{R}) \quad and \quad tq(t) \in L_2(\mathbb{R}) \tag{6.1}$$

$$\Rightarrow t^{1/2}q(t) \in L_2(\mathbb{R}) \Rightarrow q(t) \in L_1(\mathbb{R}) \Rightarrow \hat{q}(\omega) \text{ is continuous.}$$

For good time as well as frequency localization, we impose the further restriction that

$$\omega\hat{q}(\omega) \in L_2(\mathbb{R}) \tag{6.2}$$

Gabor used the Gaussian window

$$q_\alpha(t) = \frac{1}{2\sqrt{\pi\alpha}} e^{-t^2/4\alpha}$$

$$Q_\alpha(\omega) \triangleq \hat{q}_\alpha(\omega) = \frac{1}{2\sqrt{\pi\alpha}} \left(\sqrt{\frac{\pi}{1/4\alpha}} \right) e^{-\omega^2/4\alpha(1/4\alpha)} = e^{-\omega^2}$$

$$\therefore \hat{q}_\alpha(0) = \int_{-\infty}^{\infty} q_\alpha(t)dt = \int_{-\infty}^{\infty} q_\alpha(t - \beta)dt = 1, \beta \in \mathbb{R}$$

$$\text{Center:} \frac{1}{\|q_\alpha\|_2} \int_{-\infty}^{\infty} t|q_\alpha(t)|^2 dt = 0$$

Define a family $\{q_\alpha(t)\}$ of *Gaussian* window functions.

$$q_\alpha(t) = \frac{1}{2\sqrt{\pi\alpha}} e^{-t^2/4\alpha} \tag{6.3}$$

Recall the identities,

$$\int_{-\infty}^{\infty} e^{-at^2} dt = \sqrt{\frac{\pi}{a}}$$

and

$$\int_{-\infty}^{\infty} e^{-p^2t^2 \pm qt} dt = \int_{-\infty}^{\infty} e^{(-pt \pm q/2p)^2} e^{q^2/4p^2} dt = e^{q^2/4p^2} \frac{\sqrt{\pi}}{p} \tag{6.4}$$

$$\|q_\alpha\|^2 = \int_{-\infty}^{\infty} \frac{1}{4\pi\alpha} e^{-t^2/2\alpha} dt = \frac{1}{\sqrt{8\pi\alpha}} \tag{6.5}$$

the time width Δ_t is defined to be

$$(\Delta_t)^2 = \frac{1}{\|q_\alpha\|^2} \left[\int_{-\infty}^{\infty} t^2 |q_\alpha(t)|^2 dt \right] \tag{6.6}$$

Integrate by parts to show that the integral within square brackets in the above equation evaluates to

$$\frac{1}{4\pi\alpha} \int_{-\infty}^{\infty} t^2 e^{-t^2/2\alpha} dt = \sqrt{\frac{\alpha}{8\pi}} \tag{6.7}$$

Substituting the last equation in the equation preceding it, one gets

$$\Delta_t^2 = \alpha \tag{6.8}$$

The Fourier transform $Q_\alpha(\omega)$ of $q_\alpha(t)$ is

$$Q_\alpha(\omega) = \int_{-\infty}^{\infty} \frac{1}{2\sqrt{\pi\alpha}} e^{-t^2/4\alpha} e^{-j\omega t} dt$$

and applying to this the identity in (6.3)

$$Q_\alpha(\omega) = \frac{1}{2\sqrt{\pi\alpha}} e^{-\omega^2\alpha} \sqrt{4\pi\alpha} = e^{-\omega^2\alpha}$$

$$\|Q_\alpha(\omega)\|^2 = \int_{-\infty}^{\infty} e^{-2\omega^2\alpha} d\omega = \sqrt{\frac{\pi}{2\alpha}}$$

Therefore, the frequency width, Δ_ω, defined by

$$\Delta_\omega^2 = \frac{1}{\|Q_\alpha(\omega)\|^2} \int_{-\infty}^{\infty} \omega^2 |Q_\alpha(\omega)|^2 = \sqrt{\frac{2\alpha}{\pi}} \int_{-\infty}^{\infty} \omega^2 e^{-2\alpha\omega^2} d\omega \qquad (6.9)$$

But, using (6.7)

$$\int \omega^2 e^{-2\alpha\omega^2} d\omega = \frac{\pi/\alpha}{4\pi(1/4\alpha)} \int \omega^2 e^{-\frac{\omega^2}{2(1/4\alpha)}} d\omega$$

$$= \frac{\pi}{\alpha} \sqrt{\frac{1}{4\alpha} \frac{1}{8\pi}}$$

which, following the substitution in (6.9) yields

$$\Delta_\omega^2 = \frac{1}{4\alpha}.$$

Therefore,

$$\Delta_t^2 \Delta_\omega^2 = \frac{1}{4}.$$

Fact 6.2. *According to Uncertainty Principle, for functions $p(t) \in L_1 \cap L_2$*

$$\Delta_t^2 \Delta_\omega^2 \geq \frac{1}{4}$$

The above is based on the following expression for energy function E.

$$E = \int_{-\infty}^{\infty} |p(t)|^2 dt = \frac{1}{2\pi} \int_{-\infty}^{\infty} |P(\omega)|^2 d\omega.$$

6.1.6 Introduction to Continuous-Time Wavelets

The CWT has deep mathematical roots in the work of Albert P. Calderon. The Calderon identities allow one to give integral representations of many natural operators by using simple pieces of such operators. These pieces, which are essentially spectral projections, can be chosen in clever ways and have proved to be of tremendous utility in various problems of numerical analysis, multidimensional signal processing, video data compression, and reconstruction of high resolution images and high quality speech (from their degraded counterparts). The multiresolution method (in progressive transmission) decomposes high resolution images into a hierarchy of pieces or components, each more detailed than the next. Thus, the multiresolution approach, which is based on the wavelet theory of successive approximations provides the mechanism for transmitting various grades of images depending upon the quality of reconstruction sought and the limitations set by channel capacity as well as the need for compatibility with existing receivers like HDTV.

To the family of Gabor transforms $\{q_\alpha(t)\}$, introduce the time-shift parameter β to get

$$q_{\alpha,\beta}(t) = \frac{1}{2\sqrt{\pi\alpha}} e^{-(t-\beta)^2/4\alpha}$$

A property not possessed by the window function $q_{\alpha,\beta}(t)$, which can span any function $p(t)$ for all possible choices of the parameters α and β, is with respect to its average value, which is non-zero, i.e.

$$\int_{-\infty}^{\infty} q_{\alpha,\beta}(t)dt \neq 0$$

A dilation or scaling parameter is needed so that the window adjusts its duration according to the frequency. For a fixed window duration, when $p(t)$ is a high frequency signal, many cycles are captured by the window, whereas if $p(t)$ is a low frequency signal, only very few cycles are within the same window. Thus, the accuracy of the estimate of the Fourier transform is poor at low frequencies and improves as the frequency increases. Other limitations of the Gabor transform are that it does not give rise to an orthonormal signal representation and does not provide a stable reconstruction. Consider, next the window function of the form,

$$w_{\alpha,\beta}^a(t) = \alpha^{-a/2} w(\alpha^{-a}(t-\beta)), \quad a > 0, \alpha \in \mathbb{R}^+, \beta \in \mathbb{R}$$

The Fourier transform of $w_{\alpha,\beta}^a$ is

$$W_{\alpha,\beta}^a(\omega) = \alpha^{a/2} e^{j\beta\omega} W(\alpha^a \omega)$$

where $W(\omega)$ is the Fourier transform of $w(t)$. In particular, when $a = 1$, then $w_{\alpha,\beta}^1(t)$ will be long and of low frequency for $\alpha > 1$ and will be of short and high frequency for $\alpha < 1$. Suppose that $w(t) \in L^2$ is centered at $t = t_0$ and its "time-width" is Δ_t. Then, the window function, $w_{\alpha,\beta}^1(t)$ has center at $\beta + \alpha t_0$ and its time-width is $\alpha\Delta_t$. Suppose that $W(\omega)$ is centered at $\omega = \omega_0 > 0$ and its frequency width is Δ_ω. Then, the window function and, consequently, the wavelet transform

$$WT_p(\alpha,\beta) = \int_{-\infty}^{\infty} p(\tau)w_{\alpha,\beta}^1(\tau)d\tau = \int_{-\infty}^{\infty} p(\tau)w\left(\frac{t-\beta}{\alpha}\right)d\tau$$

provides local information in the frequency domain over the interval $\left[\frac{\omega_0}{\alpha} - \frac{\Delta_\omega}{2\alpha}, \frac{\omega_0}{\alpha} + \frac{\Delta_\omega}{2\alpha}\right]$. The quality factor Q of the bandpass filter, then, is

$$Q = \frac{\text{Center Frequency}}{\text{Bandwidth}} = \frac{\omega_0/\alpha}{2\Delta\omega/2\alpha} = \frac{\omega_0}{\Delta_\omega}$$

which is independent of the scaling factor α. Thus, the sequence of bandpass filters are constant-Q filters. The inverse wavelet transform is

$$p(t) = \frac{1}{C_W} \int_{-\infty}^{\infty} \int_{-\infty}^{\infty} WT_p(\alpha, \beta) w_{\alpha,\beta}(t) \frac{d\alpha}{\alpha^2} d\beta$$

where $\alpha > 0$, where the admissibility constant C_W satisfies

$$C_W = \int_{-\infty}^{\infty} \frac{|W(\omega)|^2}{|\omega|} d\omega < \infty$$

6.1.7 Generalized Parseval and Resolution of Identity in CWT

Mother Wavelet $\psi(t) \in L_2(\mathbb{R})$ and, for rigour in the derivations it is also assumed that $\psi(t) \in L_1(\mathbb{R})$ (the former assumption allows a Hilbert space and the latter assumption makes the Fourier transform of $\psi(t)$ continuous and permits interchanging the order of integration).

Scaled and dilated version for $\alpha \in R \backslash \{0\}, \beta \in \mathbb{R}, 2m \in \mathbb{R}^+ \bigcup \{0\}$, is

$$\psi_{\alpha,\beta}(t) = |\alpha|^{-m} \psi\left(\frac{t - \beta}{\alpha}\right) \tag{6.10}$$

$$\|\psi_{\alpha,\beta}(t)\|^2 = |\alpha|^{-2m} \int_{-\infty}^{\infty} |\psi\left(\frac{t - \beta}{\alpha}\right)|^2 \, dt = |\alpha|^{1-2m} \|\psi(t)\|^2 \tag{6.11}$$

$$\therefore \psi_{\alpha,\beta}(t) \in L_2(\mathbb{R}).$$

The wavelet $\psi(t)$ must satisfy the *admissibility* condition, i.e.

$$C_\psi = \int_{-\infty}^{\infty} \frac{|\Psi(\omega)|^2}{|\omega|} \, d\omega < \infty, \Psi(\omega) \triangleq \hat{\psi}(\omega) = FT(\psi(t)).$$

The above condition is satisfied by forcing $\Psi(\omega)$ to have sufficient decay as $|\omega| \to \infty$ *and*

$$\hat{\psi}(0) = \int_{-\infty}^{\infty} \psi(t) \, dt = 0.$$

It is convenient, as noted above, to take $\psi(t) \in L_1(\mathbb{R}) \bigcap L_2(\mathbb{R})$. $\psi(t)$, which is oscillatory by previous equation, behaves as the impulse response of a bandpass filter that decays. The continuous wavelet transform (CWT),

$$CWT_f(\alpha, \beta) : L_2(\mathbb{R}) \longrightarrow L_2(\mathbb{R}^2)$$

of a function $f(t) \in L_2(\mathbb{R})$ onto a time-scale space $L_2(\mathbb{R}^2)$ is given by the *ANALYSIS FORMULA*

$$\text{CWT}_f(\alpha, \beta) = \langle \psi_{\alpha,\beta}(t), f \rangle \triangleq |\alpha|^{-m} \int_{\mathbb{R}} \psi^* \left(\frac{t - \beta}{\alpha} \right) f(t) \, dt \qquad (6.12)$$

The *RESOLUTION OF IDENTITY* gives the *SYNTHESIS FORMULA*

$$f(t) = \frac{1}{C_\psi} \int_{-\infty}^{\infty} \int_{-\infty}^{\infty} \text{CWT}_f(\alpha, \beta) \psi_{\alpha,\beta}(t) \frac{d\alpha d\beta}{\alpha^2} \qquad (6.13)$$

Henceforth, WLOG, take $m = 1/2$ and recall then

$$\frac{1}{2\pi} \hat{\psi}_{\alpha,\beta}(\omega) = \frac{|\alpha|^{-1/2}}{2\pi} \int_{-\infty}^{\infty} \psi \left(\frac{t - \beta}{\alpha} \right) e^{-j\omega t} \, dt.$$

$$= \frac{\alpha |\alpha|^{-1/2}}{2\pi} e^{-j\beta\omega} \hat{\psi}(\alpha\omega) \qquad (6.14)$$

WLOG, restrict attention to *real-valued wavelets*. Then, from standard Fourier transform theory (generalized Parseval),

$$\text{CWT}_f(\alpha, \beta) = \int_{-\infty}^{\infty} f(t) \psi_{\alpha,\beta}(t) \, dt = \frac{1}{2\pi} \int_{-\infty}^{\infty} \hat{f}(\omega) \psi_{\alpha,\beta}^*(\omega) \, d\omega \qquad (6.15)$$

Equation (6.15) follows after noting that

$$\text{CWT}_f(\alpha, \beta) = \left[f(t) * \psi_{\alpha,\beta}(-t) \right]_{t=\beta}$$

But, from Eq. (6.14)

$$\hat{\psi}_{\alpha,\beta}(\omega) = \alpha |\alpha|^{-1/2} e^{-j\beta\omega} \hat{\psi}(\alpha\omega) \qquad (6.16)$$

From (6.16) and (6.15),

$$\text{CWT}_f(\alpha, \beta) = \frac{1}{2\pi} \left[\int_{-\infty}^{\infty} \hat{f}(\omega) \hat{\psi}^*(\alpha\omega_1) e^{j\beta\omega_1} \, d\omega_1 \right] \frac{\alpha}{|\alpha|^{1/2}} \qquad (6.17)$$

For another function $g(t) \in L_2(\mathbb{R})$

$$\text{CWT}_g(\alpha, \beta) = \frac{1}{2\pi} \left[\int_{-\infty}^{\infty} \hat{g}(\omega_1) \hat{\psi}^*(\alpha\omega_1) e^{j\beta\omega_1} \, d\omega_1 \right] \frac{\alpha}{|\alpha|^{1/2}} \qquad (6.18)$$

From (6.17) and (6.18),

$$\therefore I \triangleq \int_{-\infty}^{\infty} \int_{-\infty}^{\infty} \mathrm{CWT}_f(\alpha, \beta) \mathrm{CWT}_g^*(\alpha, \beta) \frac{d\alpha d\beta}{\alpha^2}$$

$$= \frac{1}{(2\pi)^2} \left[\int_{-\infty}^{\infty} \int_{-\infty}^{\infty} d\alpha d\beta \right.$$

$$\left. \int_{-\infty}^{\infty} \int_{-\infty}^{\infty} d\omega d\omega_1 \, \hat{g}^*(\omega_1) \hat{f}(\omega) \hat{\psi}^*(\alpha\omega) \hat{\psi}(\alpha\omega_1) e^{j\beta(\omega-\omega_1)} \right] \frac{1}{|\alpha|}$$

Define,

$$H(\omega, \omega_1, \alpha) \triangleq \hat{g}^*(\omega_1) \hat{f}(\omega) \hat{\psi}^*(\alpha\omega) \hat{\psi}(\alpha\omega_1)$$

Interchanging the order of integration (Fubini's Theorem),

$$I = \frac{1}{(2\pi)^2} \left[\int_{-\infty}^{\infty} d\alpha \int_{-\infty}^{\infty} \int_{-\infty}^{\infty} d\omega d\omega_1 \, H(\omega, \omega_1, \alpha) \int_{-\infty}^{\infty} e^{j\beta(\omega-\omega_1)} \, d\beta \right] \frac{1}{|\alpha|}$$

The problem of reducing double (or multiple) integrals plays an important role in classical analysis. In the Lebesgue integration theory, the key result along these lines is Fubini's Theorem, basic in the theory of multiple integration, which asserts that if the double integral exists then so do the iterated integrals and, moreover, that they are equal.
But,

$$\lim_{B \to \infty} \int_{-B}^{B} e^{j\beta(\omega-\omega 1)} d\beta = \lim_{B \to \infty} 2\pi \frac{\sin(\omega-\omega_1)B}{\pi(\omega-\omega_1)} = 2\pi\delta(\omega-\omega_1)$$

$$\therefore I = \frac{1}{(2\pi)} \left[\int_{-\infty}^{\infty} d\alpha \int_{-\infty}^{\infty} \int_{-\infty}^{\infty} d\omega d\omega_1 \, H(\omega, \omega_1, \alpha) \delta(\omega-\omega_1) \right] \frac{1}{|\alpha|}$$

$$= \frac{1}{(2\pi)} \int_{-\infty}^{\infty} \frac{d\alpha}{|\alpha|} \int_{-\infty}^{\infty} d\omega \, \hat{g}^*(\omega) \hat{f}(\omega) |\Psi(\alpha\omega)|^2$$

$$= \int_{-\infty}^{\infty} \frac{|\Psi(x)|^2}{|x|} dx \frac{1}{2\pi} \int_{-\infty}^{\infty} \hat{g}^*(\omega) \hat{f}(\omega) d\omega, \quad x = \alpha\omega$$

$$= C_\psi \langle f(t), g(t) \rangle.$$

In the above expression whenever $f = g$ the function $C_\psi(f, g)$ defines the norm of $f(org)$ in the Hilbert space of the space CWT for $f(t) \in L_2(R)$. Note that this Hilbert space is also an RKHS.

6.1.8   The Derivation of the ICWT

Here, we present a shorter proof based on the fact given next.

Fact 6.3.

$$\langle g(t), f(t) \rangle = \langle f(t), g(t) \rangle = \frac{1}{C_\psi} \int_{-\infty}^{\infty} \int_{-\infty}^{\infty} CWT_f(\alpha, \beta) CWT_g^*(\alpha, \beta) \, \frac{d\alpha d\beta}{\alpha^2}$$

Choose

$$g_\gamma(t) = \frac{1}{2\sqrt{\pi \alpha}} e^{-t^2/4\gamma}, \quad \gamma > 0$$

$$g_\gamma(t) \overset{FT}{\longleftrightarrow} e^{-\gamma \omega^2}$$

It can be strictly proved with mild restrictions on $f(t)$: ($f(t)$ is continuous and $f(t) \in L^1$ suffices) that

$$\lim_{\gamma \to 0^+} \int_{-\infty}^{\infty} f(t-x) g_\gamma(x) \, dx \to f(t-0) = f(t)$$

Consider now the limit of the RHS of the equation in the preceding Fact as given in the above equation

$$\frac{1}{C_\psi} \lim_{\gamma \to 0^+} \int_{-\infty}^{\infty} \int_{-\infty}^{\infty} CWT_f(\alpha, \beta) \langle \psi_{\alpha,\beta}(x), g_\gamma(t-x) \rangle^* \, \frac{d\alpha d\beta}{\alpha^2}$$

$$= \frac{1}{C_\psi} \int_{-\infty}^{\infty} \int_{-\infty}^{\infty} CWT_f(\alpha, \beta) \psi_{\alpha,\beta}(t) \, \frac{d\alpha d\beta}{\alpha^2}$$

$$= \lim_{\gamma \to 0} \langle f(t), g_\gamma(t-x) \rangle$$

$$= f(t)$$

Thus, the inverse CWT is obtained.

6.1.9   Noble Identities

For filter bank implementation to generate discrete-time wavelets and for perfect reconstruction from analysis and synthesis bank of filters, downsampling, upsampling, filtering are required. Interchange of filtering and downsampling operations, as well as upsampling and filtering operations are crucial in efficient implementations (e.g. polyphase filter banks to be described later). Here, simple proofs are advanced to prove such identities, referred to as Noble identities.

Consider a special case when the unit impulse response sequence $\{h[n]\}$ generated by a rational, possibly non-causal, transfer function $H(z)$ has $(N-1)$ zeros between each pair of nonzeros. Then, there exists a filter unit impulse response $\{g[n]\}$ such that $g[n] = h[nN], \forall n$, and $H(z) = G(z^N)$.

6.1.9.1 First Noble Identity

$$G(z)(\downarrow N) = (\downarrow N)G(z^N)$$

Proof. Downsampling by positive integer N converts the input sequence $x[k]$ to $x[kN]$. Then, the filter $G(z)$ produces, for the left hand side,

$$v[l] = \sum_k g[k]x[N(l-k)]$$

Since $H(z) = G(z^N)$ uses the Nth sample of the input, therefore this filter generates,

$$v[l] = \sum_k g[k]x[l-Nk]$$

In the right hand side, this filtering operation is followed by downsampling which replaces l by Nl to give the same output as in the first case (left hand side). The first noble identity allows decimation to precede filtering, because one is tempted to filter the input signal first by an anti-aliasing filter to avoid potential aliasing due to decimation. The goal for interchange of expander and interpolating filter is to perform filtering before upsampling (more efficient due to reduced number of filtering operations). Consider the special case when the interpolating filter $F(z) \leftrightarrow f[n]$ has $(N-1)$ zero coefficients between each pair of non zeros. Then, there exists a filter with unit impulse response $g[n] = f[nN], \forall N$ i.e. $F(z) = G(z^N)$

6.1.9.2 Second Noble Identity

$$(\uparrow N)G(z) = G(z^N)(\uparrow N)$$

It is easy to prove after noting that the second Noble identity is the dual of the first Noble identity. We state without proof another commutativity result.

Fact 6.4. *Given integers $M \geq 2, N \geq 2$, the identity $(\uparrow M)(\downarrow N) = (\downarrow N)(\uparrow M)$ holds if and only if M and N are relatively prime. In that case, $V = (\uparrow M)(\downarrow N)$ has components*

$$u[k] = \begin{cases} x[kN], \text{ if } k \text{ is divisible by } M \\ 0, \text{ otherwise.} \end{cases}$$

6.1.10   Upsampling and Downsampling Operations in z-Domain

Given $v = (\downarrow 2)x$ and $u = (\uparrow 2)v$, the z-transform of v and u are

$$V(z) = \frac{1}{2}\left[X(z^{1/2}) + X(-z^{1/2})\right]$$

$$U(z) = \frac{1}{2}\left[X(z) + X(-z)\right]$$

Proof.

$$u = (\uparrow 2)v \leftrightarrow \begin{cases} u[2k] = v[k] \\ u[2k+1] = 0 \end{cases}$$

Therefore,

$$U(z) = \sum u[k]z^{-k} = \sum u[2k]z^{-2k}$$

$$= \sum v[k]z^{-2k} = V(z^2)$$

$$v = (\downarrow 2)x \leftrightarrow v[k] = x[2k]$$

Form an auxiliary sequence,

$$\{w[k]\} = (\ldots x[0]\ 0\ x[2]\ 0 \ldots)$$

Then

$$(\downarrow 2)x = (\downarrow 2)w$$

$$W(z) = \frac{1}{2}\sum_k x[k]z^{-k} + \frac{1}{2}\sum_k (-1)^k x[k]z^{-k}$$

$$= \frac{1}{2}[X(z) + X(-z)]$$

$$V(z) = \sum_k x[2k]z^{-k}$$

since,

$$v = (\downarrow 2)x \leftrightarrow v[k] = x[2k]$$

Furthermore, from the construction of $w[k]$, it follows that

$$V(z) = \sum_k w[2k]z^{-2k} = \sum_k w[k]z^{-k/2} = W(z^{1/2})$$

$$= \frac{1}{2}[X(z^{1/2}) + X(-z^{1/2})]$$

$$U(z) = V(z^2) = \frac{1}{2}[X(z) + X(-z)]$$

Generalization Given $v = (\downarrow N)x$ and $u = (\uparrow N)v$, the z-transforms of v and u are

$$V(z) = \frac{1}{N} \sum_{k=0}^{N-1} X\left(z^{1/N} e^{j2\pi k/N}\right)$$

$$U(z) = \frac{1}{N} \sum_{k=0}^{N-1} X\left(z e^{j2\pi k/N}\right)$$

6.1.11 General LP and Orthonormal PR QMF Filter Banks with Lattice Realization

First, we start with LP and then proceed to orthonormal filter banks. For both $H_0(z)$ and $H_1(z)$ to have the same lengths (EVEN) and be linear phase, one, say $H_0(z)$ must be symmetric while the other must be anti-symmetric. Let L be the common filter length. Then,

$$H_0(z) = z^{-(L-1)} H_0(z^{-1}), H_1(z) = -z^{-(L-1)} H_1(z^{-1})$$

6.1.11.1 Polyphase Decompositions

$$H_0(z) = H_{00}(z^2) + z^{-1} H_{01}(z^2) = z^{-(L-1)} H_0(z^{-1})$$

$$= z^{-(L-1)} \left[H_{00}(z^{-2}) + z H_{01}(z^{-2})\right]$$

$$= z^{-(L-2)} \left[H_{01}(z^{-2}) + z^{-1} H_{00}(z^{-2})\right]$$

$$H_1(z) = H_{10}(z^2) + z^{-1} H_{11}(z^2) = z^{-(L-1)} H_1(z^{-1})$$

$$= -z^{-(L-1)} \left[H_{10}(z^{-2}) + z H_{11}(z^{-2})\right]$$

$$= -z^{-(L-2)} \left[H_{11}(z^{-2}) + z^{-1} H_{10}(z^{-2})\right]$$

The *Polyphase Matrix*, then, becomes

$$H_p(z) = \begin{bmatrix} z^{-L/2+1}H_{01}(z^{-1}) & z^{-L/2+1}H_{00}(z^{-1}) \\ -z^{-L/2+1}H_{11}(z^{-1}) & -z^{-L/2+1}H_{10}(z^{-1}) \end{bmatrix}$$

$$= z^{-L/2+1} \begin{bmatrix} 1 & 0 \\ 0 & -1 \end{bmatrix} \begin{bmatrix} H_{01}(z^{-1}) & H_{00}(z^{-1}) \\ H_{11}(z^{-1}) & H_{10}(z^{-1}) \end{bmatrix}$$

$$= z^{-L/2+1} \begin{bmatrix} 1 & 0 \\ 0 & -1 \end{bmatrix} \begin{bmatrix} H_{00}(z^{-1}) & H_{01}(z^{-1}) \\ H_{10}(z^{-1}) & H_{11}(z^{-1}) \end{bmatrix} \begin{bmatrix} 0 & 1 \\ 1 & 0 \end{bmatrix}$$

We focus the realization aspects with respect to orthonormal filter banks using paraunitariness. The synthesis filters satisfy

$$G_0(z) = H_1(-z), \ G_1(z) = -H_0(-z)$$

Thus, alias cancellation, linear phase, and also excellent attenuation (plus the convenience of VLSI implementable lattice realization) is obtained.

We now work towards a canonic structure for orthonormal FIR 2-channel filter bank that has a lattice implementation and is highly modular. We have seen that 2-channel PR linear phase FIR filter banks that are also orthogonal are not possible if the length exceeds 2. To construct orthogonal filters, the paraunitary condition must be satisfied by the polyphase matrices in the filter bank. To wit, for the polyphase analysis matrix (we restrict attention, without loss of generality, to the real coefficients case).

$$H_p(z)H_p^t(z^{-1}) = H_p^t(z^{-1})H_p(z) = 1$$

A paraunitary matrix is unitary on $|z| = 1$. Recall that a matrix U is unitary if

$$UU^H = U^H U = I$$

where the superscript H denotes a "complex conjugate transpose" or "Hermitian conjugate". An orthonormal matrix is a real valued unitary matrix. The basic building blocks of a paraunitary matrix of order 2 are the rotation matrix and the delay matrix. These simple building blocks, each having a lattice realization, are cascaded and the result is a paraunitary matrix of the higher order that has a lattice realization. To wit, a $H_p^l(z)$ of degree l is related to $H_p^{l-1}(z)$ of degree $l-1$ generically by

$$H_p^l(z) = \begin{bmatrix} \cos\theta_l & -\sin\theta_l \\ \sin\theta_l & \cos\theta_l \end{bmatrix} \begin{bmatrix} 1 & 0 \\ 0 & z^{-1} \end{bmatrix} H_p^{l-1}(z)$$

Analysis is easy, but even synthesis of $H_p^l(z)$ is possible by adaptation of "Richards' theorem" (N.K. Bose, "*Digital Filters*", 1985, pp. 227–231).

Suppose that the analysis and synthesis filters in the orthonormal filter bank are each of order $N = 2J + 1$. Then, the step-down algorithm proceeds as follows:

Algorithm 6.1. *Step-down Algorithm*

(a) ***Initialization:*** $H_0^J(z) \triangleq H_0(z)$ *Alternating flip* $\Rightarrow H_1^J(z) = H_1(z) = -z^{-(2J+1)}H_0(-z^{-1})$

(b) ***Stepdown:*** *For* $m = J, J - 1, \ldots$

$$(1 + \alpha_m^2)H_0^{(m-1)}(z) = H_0^{(m)}(z) - \alpha_m H_1^{(m)}(z)$$

$$(1 + \alpha_m^2)z^{-2}H_1^{(m-1)}(z) = \alpha_m H_0^{(m)}(z) + H_1^{(m)}(z)$$

$$\therefore \frac{H_1^{(m-1)}(z)}{H_0^{(m-1)}(z)} = z^2 \frac{\alpha_m - \left(-\frac{H_1^{(m)}(z)}{H_0^{(m)}(z)}\right)}{\left(-\alpha_m \frac{H_1^{(m)}(z)}{H_0^{(m)}(z)}\right) + 1}$$

Example 6.1. Let us design a Daubechies maximally flat filter D_2.

$$P(z) = H_0(z)H_0(z^{-1}) = (1 + z^{-1})^k(1 + z)^k R(z),$$

where,

$$R(z) = R(z^{-1}), \quad R(e^{j\omega}) \geq 0,$$

$$R(z) = \sum_{l=-(k-1)}^{k-1} r(l)z^{-l}, \quad r(l) = r(-l)$$

Here, where $k = 2$, $R(z)$ is of the form

$$R(z) = az + b + az^{-1}$$

Since $P(z) + P(-z) = 2$, it follows that

$$R(z) = -\frac{1}{16}z + \frac{1}{4} - \frac{1}{16}z^{-1}, \quad R(e^{j\omega}) \geq 0, \forall \omega$$

After spectral factorization

$$H_0(z) = \frac{1}{4\sqrt{2}}\left[(1 + \sqrt{3}) + (3 + \sqrt{3})z^{-1} + (3 - \sqrt{3})z^{-2} + (1 - \sqrt{3})z^{-3}\right]$$

$$h_0[n] :\rightarrow \frac{1}{4\sqrt{2}} \left[1 + \sqrt{3}\ 3 + \sqrt{3}\ 3 - \sqrt{3}\ 1 - \sqrt{3}\ 0\ 0 \right]$$

$$h_0[n-2] :\rightarrow \frac{1}{4\sqrt{2}} \left[0\ 0\ 1 + \sqrt{3}\ 3 + \sqrt{3}\ 3 - \sqrt{3}\ 1 - \sqrt{3} \right]$$

$$\langle h_0[n], h_0[n-2] \rangle = (3 - \sqrt{3})(1 + \sqrt{3}) + (3 + \sqrt{3})(1 - \sqrt{3}) = 0.$$

6.1.12 Projection Operator-Based Approach

$M_m(K)$ denotes the space of square matrices of size m, taking its coefficients over a field K. Likewise, $M_{m\ell}(K)$ is the space of m-by-ℓ rectangular matrices. Here, the field K will denote either $\mathbb{R}$ or $\mathbb{C}$. Note that the coefficients of the polynomial matrix elements may belong to any arbitrary but fixed field like a finite field, which is not of characteristic zero.

Let M be a singular matrix of $M_m(K)$ (or not of full rank in $M_{m\ell}(K)$) of rank $r < m$. The image of M forms a subspace W of K^m. There is a bounded linear map P satisfying $P^2 = P$ (idempotent) from K^m onto W and P is called a projection operator. The columns of P are the projections of the standard basis vectors and W is the image of P. Form $T = MM^*$, where M^* is the Hermitian conjugate of M. Note that T is a self-adjoint operator and $Im(T) = Im(M)$. Now, define the characteristic polynomial associated with T, as $\chi_T(x)$. M, and hence T, being of rank r implies that $\chi_T(x) = x^{(m-r)}Q(x)$, where $Q(0) \neq 0$. Then $\frac{Q(T)}{Q(0)}$ is the projection operator on T's kernel and $P = I - \frac{Q(T)}{Q(0)}$ is the projection operator on T's image. A square matrix P is an orthogonal projection matrix iff $P^2 = P = P^*$, where the complex conjugate transpose P^* of P denotes the adjoint of the matrix P. In particular, for all matrices A and B such that $Im(A) \subseteq Im(M)$ and $Im(B) \subseteq Im(M)^\perp$

$$PA = A \quad , \quad PB = 0$$

Let $A(z^{-1})$ be a $(m \times m)$ paraunitary polynomial matrix of exact degree d (i.e. $A_d \neq 0$) in:

$$A(z^{-1}) = \sum_{i=0}^{d} z^{-i} A_i \quad , \quad A^*(\bar{z}^{-1}) = \sum_{i=0}^{d} z^{-i} A_i^*$$

The paraunitary property yields $A^*(\bar{z}^{-1})A(z) = I$, from which we can derive a system of quadratic equations in terms of the coefficient matrices. One of these equations is $A_0^* A_d = 0$. Since $A_d \neq 0$, therefore $rank(A_0) < m$ and $A_0 \perp A_d$. This enables ones to define a projection matrix P associated with A_0, verifying also that $PA_d = 0$.

Theorem 6.1. *If P is the projection matrix associated with A_0, then $P + (I-P)z^{-1}$ is a paraunitary left factor of the paraunitary matrix $A(z^{-1})$.*

Proof. The paraunitariness of $P + (I - P)z^{-1}$ follows from:

$$(P + (I - P)\bar{z}^{-1})^*(P + (I - P)z)$$

$$= (P^* + (I - P^*)z^{-1})(P + (I - P)z)$$

$$= P^2 + Pz - P^2z + Pz^{-1} - P^2z^{-1} + I - 2P + P^2$$

$$= P + Pz - Pz + Pz^{-1} - Pz^{-1} + I - 2P + P$$

$$= I$$

The inverse of this paraunitary factor is evidently $P + (I - P)z$.

Furthermore, $P + (I - P)z^{-1}$ is a left factor of $A(z^{-1})$ as justified next. Let $A(z^{-1}) = A_0 + A_1 z^{-1} + \ldots + A_d z^{-d}$. Then we can factor $P + (I - P)z^{-1}$ and compute the remainder by left-multiplying $A(z^{-1})$ with the inverse factor $P + (I - P)z$.

$$(P + (I - P)z)A(z^{-1})$$

$$= PA_0 + PA_1 z^{-1} + \ldots + PA_d z^{-d}$$

$$+ A_0 z + A_1 + \ldots + A_d z^{-d+1} -$$

$$- PA_0 z - PA_1 - \ldots - PA_d z^{-d+1}$$

Since $PA_0 = A_0$, we can compute the term with positive power of z, as

$$A_0 z - PA_0 z = 0.$$

Moreover since $PA_d = 0$, therefore the remainder is a polynomial matrix in z^{-1} of reduced degree $d - 1$. Note that the computation of a factor in this approach depends only on the constant term of the matrix.

Theorem 6.2. *Let $A(z^{-1})$ be a $m \times m$ paraunitary matrix with polynomial degree one, and A_0 its constant coefficient. Then*

$$rank(A_0) = r \Leftrightarrow det(A(z^{-1})) = \pm z^{-(m-r)}$$

Proof. A being of polynomial degree one, apply the factorization algorithm.

$$A(z^{-1}) = [P + (I - P)z^{-1}]R$$

then,

$$R = \underbrace{PA_0}_{A_0} + \underbrace{(I - P)A_0}_{0} z + \underbrace{PA_1}_{0} z^{-1} + \underbrace{(I - P)A_1}_{A_1}$$

$$R = A_0 + A_1 = A(1)$$

Define $L = P + (I - P)z^{-1}$, then $det(A(z^{-1})) = \pm det(L)$, since $A(1)$ is unitary. Also note that $A_0 = PA(1)$ and since $A(1)$ is non-singular, $rank(A_0) = rank(P)$.

Lastly, since P is an orthogonal projection operator, there exists a matrix Q, non-singular, such that, if $rank(P) = r$,

$$P = Q^{-1} \begin{pmatrix} 1 & & & & \\ & \ddots & & & \\ & & 1 & & \\ & & & 0 & \\ & & & & \ddots \\ & & & & & 0 \end{pmatrix} Q$$

with r ones on the diagonal of the central matrix. Since $I = Q^{-1}Q$, therefore,

$$L = P + (I - P)z^{-1}$$

$$= Q^{-1} \begin{pmatrix} 1 & & & & \\ & \ddots & & & \\ & & 1 & & \\ & & & z^{-1} & \\ & & & & \ddots \\ & & & & & z^{-1} \end{pmatrix} Q$$

Since $det(L) = \pm z^{-(m-r)}$, the following equivalence is obtained:

$$det(A) = \pm z^{-(m-r)} \Longleftrightarrow det(L) = \pm z^{-(m-r)}$$

$$\Longleftrightarrow rank(P) = r \Longleftrightarrow rank(A_0) = r$$

The algorithm, summarized in the next subsection, works with sequential reduction of polynomial degree though the determinantal degree may reduce by more than 1 (determinantal degree reduces by 1 when rank $P \triangleq r$ equals $m - 1$). When the projection matrix is of lower rank ($< m - 1$), further determinantal factorization would still be possible to arrive at the atomic factorization.

Note: The product of two atomic factors is a factor, of determinantal degree two, and, in the general case, of polynomial degree one. This later case corresponds to a low-rank constant coefficient matrix. It is also a case when factorization is very simple.

In general, low-rank constant coefficient matrices correspond to matrices whose columns and rows can be rearranged into a block-diagonal matrix. Then a first step to factorization is to separate those blocks by a simple factorization. Also note that after this factorization, the constant coefficient matrix of the remainder will be of rank $m - 1$.

6.1.13 Algorithm in m-Channel Case

The main interest of the projection algorithm is the simplicity of computation it offers in case of matrices of higher dimensions. The algorithm can be written in a few simple steps.

For an m-by-m matrix $A(z^{-1}) \triangleq A$, the algorithm is described by the steps below.

Algorithm 6.2. *Initialize* $A_{(1)} = A$. *Then, until* $A_{(k)}$ *becomes an 'elementary' factor, do the following steps for* $k = 1, 2, \ldots$

(a) *Define by* $M_{(k)}$ *the constant coefficient matrix of* $A_{(k)}$.
(b) *Define* $T_{(k)} \triangleq M_{(k)} M_{(k)}^*$ *and* $\chi_T(x) = det(T_{(k)} - xI)$.
(c) *Define* $r = rank(M_{(k)})(= rank(T_{(k)}))$, *then form* $Q_{(k)}(x) = x^{(r-m)} \chi_T(x)$
(d) $P_{(k)} = I - \frac{Q_{(k)}(T)}{Q_{(k)}(0)}$.
(e) $L_{(k)}(z^{-1}) = P_{(k)} + (I_m - P_{(k)})z^{-1}$
(f) $R_{(k)}(z^{-1}) = L_{(k)}(z)A_{(k)}(z^{-1})$
(g) $A_{(k+1)} = R_{(k)}(z^{-1})$
(h) • *If order (polynomial degree) of* $R_{(k)}(z^{-1})$ *is 1 then stop*
    • *Else go to step 1*

After the algorithm terminates, the factorization is given by $A = L_{(1)} \ldots L_{(k)} R_{(k)}$

6.1.13.1 Examples

Example 6.2. In the two-channel case, the computation of the projection matrix is simplified to be:

$$P = \frac{A_0 A_0^*}{Tr(A_0 A_0^*)},$$

which has Hermitian symmetry and is, equivalently, orthogonal. Indeed the characteristic polynomial becomes: $\chi_T(x) = x(x - Tr(T))$ and then $Q(T) = T - Tr(T) \cdot I$, $Q(0) = -Tr(T)$,

$$P = I - \frac{Q(T)}{Q(0)} = \frac{T}{Tr(T)}$$

Note that in the two-channel case, A_0 is always of rank 1, which since $m = 2$ is also $m-1$ (where m is the order of the matrix). Therefore, there is no 'low-rank case' contrary to what could take place in the general case. Hence the polynomial degree (order) of the matrix is equal to the degree of the determinant and also exactly equal to the number of factors we can obtain. Therefore this factorization is coherent with the result obtained by Vaidynathan and Hoang for the 2-channel case.

Example 6.3. Here is an example for 3×3 matrices i.e. $m = 3$:

$$A(z^{-1}) \triangleq A = \begin{pmatrix} \frac{1-z^{-1}}{2} & \frac{z^{-1}-z^{-3}}{4} & \frac{z^{-1}+2z^{-2}+z^{-3}}{4} \\ \frac{1+z^{-1}}{2} & \frac{z^{-1}-2z^{-2}+z^{-3}}{4} & \frac{z^{-1}-z^{-3}}{4} \\ 0 & \frac{1+z^{-1}}{2} & \frac{1-z^{-1}}{2} \end{pmatrix}$$

Let $P_{(k)}, L_{(k)}$ and $R_{(k)}$ denote respectively, the projection matrix, the associated left-factor and remainder at the kth iteration. Then, from $M_{(1)}$, the constant matrix in A, form $T_{(1)} = M_{(1)}M_{(1)}^*$.

$$T_{(1)} = \frac{1}{4} \begin{pmatrix} 1 & 0 & 0 \\ 1 & 0 & 0 \\ 0 & 1 & 1 \end{pmatrix} \begin{pmatrix} 1 & 0 & 0 \\ 1 & 0 & 0 \\ 0 & 1 & 1 \end{pmatrix}^*$$

$$= \frac{1}{4} \begin{pmatrix} 1 & 1 & 0 \\ 1 & 1 & 0 \\ 0 & 0 & 2 \end{pmatrix}$$

and clearly, $r = 2$, $\chi_T(x) = -x(x^2 - x + \frac{1}{4})$, $Q_{(1)}(T_{(1)}) = -T_{(1)}^2 + T_{(1)} + \frac{1}{4}I$

$$P_{(1)} = 4T_{(1)} - 4T_{(1)}^2 = \frac{1}{2} \begin{pmatrix} 1 & 1 & 0 \\ 1 & 1 & 0 \\ 0 & 0 & 2 \end{pmatrix}$$

$$L_{(1)} = P_{(1)} + (I_3 - P_{(1)})z^{-1}$$

$$= \frac{1}{2} \begin{pmatrix} 1 + z^{-1} & 1 - z^{-1} & 0 \\ 1 - z^{-1} & 1 + z^{-1} & 0 \\ 0 & 0 & 2 \end{pmatrix}$$

$$R_{(1)} = L_{(1)}^{-1}A = \begin{pmatrix} 0 & \frac{z^{-1}-z^{-2}}{2} & \frac{z^{-1}+z^{-2}}{2} \\ 1 & 0 & 0 \\ 0 & \frac{1+z^{-1}}{2} & \frac{1-z^{-1}}{2} \end{pmatrix}$$

We can repeat this procedure one more time and get,

$$T_{(2)} = \begin{pmatrix} 0 & 0 & 0 \\ 0 & 1 & 0 \\ 0 & 0 & \frac{1}{2} \end{pmatrix}$$

$$P_{(2)} = 3T_{(2)} - 2T_{(2)}^2 = \begin{pmatrix} 0 & 0 & 0 \\ 0 & 1 & 0 \\ 0 & 0 & 1 \end{pmatrix}$$

$$L_{(2)} = P_{(2)} + (I_3 - P_{(2)})z^{-1} = \begin{pmatrix} z^{-1} & 0 & 0 \\ 0 & 1 & 0 \\ 0 & 0 & 1 \end{pmatrix}$$

$$R_{(2)} = L_{(2)}^{-1} R_{(1)} = \begin{pmatrix} 0 & \frac{1-z^{-1}}{2} & \frac{1+z^{-1}}{2} \\ 1 & 0 & 0 \\ 0 & \frac{1+z^{-1}}{2} & \frac{1-z^{-1}}{2} \end{pmatrix}$$

After this last step, we obtain the following elementary factorization:

$$A = L_{(1)}L_{(2)}R_{(2)}$$

where the matrices on the right-hand side have been generated above.

6.2 Extension to the Multivariate Case

The approaches in algorithmic theory of polynomial ideals and modules for multivariate polynomial matrix factorization, unimodular matrix completion (and its variants) are intimately linked to the problems faced in multidimensional multiband filter bank design.

Let $z = (z_1, z_2, \ldots, z_n)$ and let $K^{q \times p}[z] = K^{q \times p}[z_1, z_2, \ldots, z_n]$ be the ring of $(q \times p)$ n-variate polynomial matrices whose elements have coefficients in the field K of complex numbers. Let $C \in K^{q \times p}[z]$ be the composite matrix $C \triangleq [A \mid B]$, of normal full rank where without loss of generality, $q \leq p$. Let $\hat{G} \in K^{q \times q}[z]$ be a *greatest left common factor (GLCF)* of A and B (assuming $\hat{G}$ exists). Note that $\hat{G}$ is restricted to be a square matrix as required in matrix-fraction descriptions of rational matrices in multidimensional system theory. For some n-variate polynomial matrices A_1 and B_1,

$$C \triangleq [A \mid B] = \hat{G}[A_1 \mid B_1]. \tag{6.19}$$

Furthermore, it is assumed that the determinant of $\hat{G}$ equals the greatest common factor (gcf) of the major determinants of C (this is needed because primitive factorization is not possible, in general, in n-D, $n > 2$ [2, pp. 64–65]). This fact is an implicit requirement for Eq. (6.20) below and is explicitly noted here for the sake of clarity. It is clear from Eq. (6.19) that all columns of C belong to the module (over the polynomial ring $K[z]$) generated by the columns of $\hat{G}$. When the *reduced minors* (minors after gcf extraction) of C have no common zero, there exists a polynomial matrix H [2] such that,

$$[A_1 \mid B_1]H = I. \tag{6.20}$$

Consequently, using Eqs. (6.19) and (6.20) one can write

$$\hat{\mathbf{G}} = \hat{\mathbf{G}}[\,\mathbf{A}_1 \mid \mathbf{B}_1\,]\mathbf{H} = \mathbf{CH}. \tag{6.21}$$

From Eq. (6.21), the columns of $\hat{\mathbf{G}}$ also belong to the module generated by the columns of $\mathbf{C}$. Therefore, the columns of $\hat{\mathbf{G}}$ and the columns of $\mathbf{C}$ generate the same module over the polynomial ring $K[\mathbf{z}]$. Let $\mathbf{G} \in K^{q \times s}[\mathbf{z}]$, $q \leq s$, have its columns formed from the Gröbner basis of the module generated by columns of $\mathbf{C}$.

6.2.1   Algorithm A: Algorithm for Multivariate Matrix Factorization

Given a normal full rank matrix $\mathbf{C} \in K^{q \times p}[\mathbf{z}]$, its GLCF, $\hat{\mathbf{G}}$ (if it exists), is computed as follows.

Algorithm 6.3. ■ **Step 1.** *Compute the set of reduced $q \times q$ minors or reduced major determinants of $\mathbf{C}$. If the reduced minors have no common zeros then proceed to Step 2; otherwise use the method in [81] to compute the GLCF in the bivariate case only.*

■ **Step 2.** *Compute a Gröbner basis for a module generated by all columns of $\mathbf{C}$ using any ordering. Denote the matrix whose columns are the elements of this Gröbner basis by $\mathbf{G} = [\mathbf{g}_1 \ \mathbf{g}_2 \ \dots \ \mathbf{g}_s]$. Select a maximal set $\{\mathbf{g}_{p(1)}, \mathbf{g}_{p(2)}, \dots, \mathbf{g}_{p(q)}\}$ of q linearly independent elements (over the field $K(\mathbf{z})$ of rational functions), where the set of subscripts $\{p(i)\}_{i=1}^q$ is a permutation of a subset of cardinality q of the set $\{1, 2, \dots, s\}$.*

■ **Step 3.** *If all columns of $\mathbf{C}$ belong to the module generated by the column vectors $\mathbf{g}_{p(1)}, \mathbf{g}_{p(2)}, \dots, \mathbf{g}_{p(q)}$, then $\hat{\mathbf{G}}$ is formed by using the maximal set of linearly independent elements of the Gröbner basis module computed in Step 2 and the algorithm is terminated. Otherwise, the algorithm cannot find a GLCF (see Example in next subsection).*

6.2.2   Remarks and Examples

Remark 6.1. The maximal set of linearly independent elements of the Gröbner basis, computed in Step 2 of Algorithm A, consists of q polynomial vectors, where q is the number of rows of $\mathbf{C}$. This will produce the right number of column vectors but these column vectors may only generate a proper submodule of the column space of $\mathbf{C}$. Therefore, this set is not necessarily a minimal generating set for the column space of $\mathbf{C}$ and, hence, may not serve as a GLCF (because the columns of a GLCF must generate the column space of $\mathbf{C}$).

Example 6.4. The reduced minors of the factorable matrix $\mathbf{C} \in K^{2 \times 3}[\mathbf{z}]$, shown below, are zero-coprime.

$$
\mathbf{C} = \begin{bmatrix} z_1 z_2^2 z_3 & 0 & -z_1^2 z_2^2 - 1 \\ z_1^2 z_3^2 + z_3 & -z_3 & -z_1^3 z_3 - z_1 \end{bmatrix} \triangleq \hat{\mathbf{G}}[\, \mathbf{A}_1 \mid \mathbf{B}_1 \,]
$$

$$
= \begin{bmatrix} -z_1^2 z_2^2 - 1 & z_1 z_2^2 z_3 \\ -z_1 & z_3 \end{bmatrix} \begin{bmatrix} z_1^3 z_2^2 z_3^2 & -z_1 z_2^2 z_3 & -z_1^4 z_2^2 z_3 + 1 \\ z_1^4 z_2^2 z_3 + z_1^2 z_3 + 1 & -z_1^2 z_2^2 - 1 & -z_1^3 (z_1^2 z_2^2 + 1) \end{bmatrix}
$$

Using degree reverse lexicographical ordering with $z_1 \succ z_2 \succ z_3$, the matrix $\mathbf{G}$ whose columns are the reduced Gröbner basis vectors of the module generated by all columns of $\mathbf{C}$ is calculated (by the program SINGULAR [82]) to be $\mathbf{G} = [\mathbf{g}_1 \ \mathbf{g}_2 \ \mathbf{g}_3]$, where $\mathbf{g}_1 = [0 \ z_3]^T, \mathbf{g}_2 = [z_3 \ 0]^T$, and $\mathbf{g}_3 = [(z_1^2 z_2^2 + 1) \ z_1]^T$. However, here the GLCF $\hat{\mathbf{G}}$ cannot be computed using Algorithm A, because no proper linearly independent subset of $\mathbf{G}$ can generate the column space of $\mathbf{C}$. Note that though the columns of $\hat{\mathbf{G}}$ generate the same module as the columns of $\mathbf{C}$, it cannot be derived from $\mathbf{G}$ by applying the algorithmic theory of Gröbner basis.

In general, the construction of $\hat{\mathbf{G}}$ will depend on the validity of a conjecture advanced in [83].[1]

Conjecture 6.1. Let d be the greatest common divisor of all major determinants of $\mathbf{C} \in K^{q \times p}[\mathbf{z}]$. If the reduced minors of $\mathbf{C}$ have no common zeros in K^n, then $\mathbf{C}$ can be factored as $\mathbf{C} = \hat{\mathbf{G}} \mathbf{C}_0$ with $\mathbf{C}_0 \in K^{q \times p}[\mathbf{z}]$ being ZLP, $\hat{\mathbf{G}} \in K^{q \times q}[\mathbf{z}]$ and det $\hat{\mathbf{G}} = d$.

The preceding conjecture is proved for the case $p = q + 1$ [83], the condition satisfied in the above example.

Definition 6.3. Let $\mathcal{N}$ be a set of integers and let a vector $\mathbf{v} = (v_1, \ldots, v_p)^T \in K^{p \times 1}[\mathbf{z}]$ for some $p \in \mathcal{N}$. Then $\mathbf{v}$ is called a *unimodular column vector* if its components generate $K[\mathbf{z}]$ i.e. if there exist $g_1, \ldots, g_p \in K[\mathbf{z}]$ such that $v_1 g_1 + \ldots + v_p g_p = 1$; a matrix $\mathbf{A} \in K^{q \times p}[\mathbf{z}]$ is called a *unimodular matrix* if its major determinants generate the unit ideal in $K[\mathbf{z}]$.

Remark 6.2. $\hat{\mathbf{G}}$ is unique upto multiplication by unimodular matrices. Hypothesize that $\mathbf{C} = \hat{\mathbf{G}}_1 [\mathbf{A}_1 \mid \mathbf{B}_1] = \hat{\mathbf{G}}_2 [\mathbf{A}_2 \mid \mathbf{B}_2]$. Then *det* $\hat{\mathbf{G}}_1$ and *det* $\hat{\mathbf{G}}_2$ each must divide the greatest common factor of the major determinants of $\mathbf{C}$. Since the sets

---

[1]In preparing this part of the book, Professor N.K. Bose either was unaware of some further development on this conjecture, or did not manage to update this part when he was suddenly passing away in 2009. In fact, the conjecture had been proved by several researchers shortly after it was posed in 1999. The additional references on this conjecture and related topics are now included in the Appendix of this chapter for the convenience of the readers while not affecting the original writing style of Professor Bose.

of major determinants of $[\mathbf{A}_1 | \mathbf{B}_1]$ and $[\mathbf{A}_2 | \mathbf{B}_2]$ are both zero-coprime, therefore $det\ \hat{\mathbf{G}}_1$, $det\ \hat{\mathbf{G}}_2$ must each be the greatest common factor of the major determinants of $\mathbf{C}$. Therefore, $det\ \hat{\mathbf{G}}_1 = \alpha\ det\ \hat{\mathbf{G}}_2$, $\alpha \in K$. Furthermore, there exists a $\mathbf{H} \in K^{p \times q}[\mathbf{z}]$ such that

$$\mathbf{CH} = \hat{\mathbf{G}}_1 [\ \mathbf{A}_1\ |\ \mathbf{B}_1\ ]\mathbf{H} = \hat{\mathbf{G}}_1 = \hat{\mathbf{G}}_2 [\mathbf{A}_2 | \mathbf{B}_2\ ]\mathbf{H} \overset{\triangle}{=} \hat{\mathbf{G}}_2 \mathbf{U}.$$

$\mathbf{U}$ must be unimodular because $det\ \hat{\mathbf{G}}_1 = \alpha\ det\ \hat{\mathbf{G}}_2$, $\alpha \in K$.

Remark 6.3. If the reduced minors of $\mathbf{C}$ have common zeros, then the algorithm will never find a GLCF even if it exists. This follows because $\hat{\mathbf{G}}$ is generated by the elements of the Gröbner basis of the module generated by the columns of $\mathbf{C}$, which implies that

$$\hat{\mathbf{G}} = \mathbf{CX}, \quad \mathbf{X} \in K^{p \times q}[\mathbf{z}].$$

Since a GLCF is assumed to exist and the reduced minors of $\mathbf{C}$ have common zeros, therefore, $\mathbf{C} = \hat{\mathbf{G}}\mathbf{C}_1$, $\mathbf{C}_1 \in K^{q \times p}[\mathbf{z}]$, where the major determinants of $\mathbf{C}_1$ have common zeros. Since $\hat{\mathbf{G}}$ is nonsingular, $\hat{\mathbf{G}} = \mathbf{CX} = \hat{\mathbf{G}}\mathbf{C}_1\mathbf{X}$ implies $\mathbf{C}_1\mathbf{X} = \mathbf{I}_q$, which contradicts the inference made earlier about the major determinants of $\mathbf{C}_1$ having common zeros.

6.2.3 Constructive Aspects of Unimodular Completion

Serre's conjecture [84] is known to be equivalent to the unimodular matrix completion question: can a row of a finite number of zero coprime n-variate polynomials belonging to the ring $K[z_1, z_2, \ldots, z_n]$ be completed to a square matrix over the same ring with a nonzero determinant in K? D. Quillen and A. Suslin [84] independently proved this conjecture. Several constructive procedures for independently verifying the following theorem was advanced later.

Theorem 6.3 (Quillen-Suslin). *Let* $\mathbf{A}$ *be a unimodular* $q \times p$ *matrix* ($q \leq p$) *over* $K[\mathbf{z}]$. *Then there exists a unimodular* $p \times p$ *matrix* $\mathbf{U}$ *over* $K[\mathbf{z}]$ *such that*

$$\mathbf{AU} = [\mathbf{I}_q \mid \mathbf{0}_{q \times (p-q)}]. \tag{6.22}$$

An heuristic algorithm based on syzygy computation which, though not universally applicable is usually computationally attractive, was advanced by Park. The main problem of the syzygy-based heuristic algorithm is the lack of an effective procedure for finding a minimal syzygy basis.

6.2.4 Algorithm for Computing a Globally Minimal Generating Matrix H

Definition 6.4. A syzygy module $Syz(\mathbf{A})$ of the $q \times p$ matrix $\mathbf{A} = [\mathbf{a}_1 \cdots \mathbf{a}_p]$, where $\mathbf{a}_1, \ldots, \mathbf{a}_p \in K^{q \times 1}[\mathbf{z}]$, is finitely generated [27, Ch.3]. Let $\mathbf{h}_1, \ldots, \mathbf{h}_s \in K^{p \times 1}[\mathbf{z}]$ form a *generating set (syzygy basis)* of $Syz(\mathbf{A})$, then the matrix $\mathbf{H} = [\ \mathbf{h}_1 \ \cdots \ \mathbf{h}_s\ ]$ is called a *generating matrix* of $Syz(\mathbf{A})$, (or the kernel representation of $\mathbf{A}$) implying that $\mathbf{AH} = \mathbf{0}$ ($\mathbf{H} = ker(\mathbf{A}) = \{\mathbf{h} \in K^{p \times 1}[\mathbf{z}] \mid \mathbf{Ah} = \mathbf{0}_{q \times 1}\}$).

In order to compute the globally minimal syzygy basis, the following propositions due to Lin [85] are required.

Proposition 6.1. *Let* $\mathbf{A} \in K^{q \times p}[\mathbf{z}]$ *be of rank* q, *with* $q < p$ *and let* $r = p - q$. *The syzygy basis,* $Syz(\mathbf{A})$ *of* $\mathbf{A}$ *has a generating matrix of minimal dimension* $p \times r, r \leq s$ *i.e. a generating matrix is* globally minimal *if and only if there exists a minor right prime (MRP)(i.e. the major determinants of* $\mathbf{H}$ *form a set of relatively prime polynomials) matrix* $\mathbf{H} \in K^{p \times r}[\mathbf{z}]$ *such that* $\mathbf{AH} = \mathbf{0}_{q \times r}$.

Proposition 6.2. *Let* $\mathbf{A} \in K^{q \times p}[\mathbf{z}]$ *be of rank* q, *with* $q < p$. $\mathbf{H}_1 \in K^{p \times s}[\mathbf{z}]$ *be a generating matrix of* $Syz(\mathbf{A})$, *with* $s > r$, *then* $Syz(\mathbf{A})$ *has a generating matrix of dimension* $p \times r$ *if and only if* $\mathbf{H}_1$ *can be factored as* $\mathbf{H}_1 = \mathbf{HE}$ *for some* $\mathbf{H} \in K^{p \times r}[\mathbf{z}], \mathbf{E} \in K^{r \times s}[\mathbf{z}]$ *with* $\mathbf{H}$ *being MRP.*

Algorithm 6.4. *Algorithm for computing* $\mathbf{H}$ *of* $Syz(\mathbf{A})$, $\mathbf{A} \in K^{p \times q}[\mathbf{z}]$.

- **Step 1.** *Compute a set of syzygy bases of* $\mathbf{A}$, *using the algorithm in [27, Ch.3]. Denote the corresponding generating matrix by* $\mathbf{H}_1 \in K^{p \times s}[\mathbf{z}]$. *If* $s = p - q$ *then* $\mathbf{H} = \mathbf{H}_1$, *otherwise proceed to the next step.*
- **Step 2.** *If possible, eliminate the columns of* $\mathbf{H}_1$ *which are linearly dependent (with coefficients in* $K[\mathbf{z}]$), *so that the remaining columns are linearly independent. Let* $\mathbf{H}_2$ *denote the matrix formed by these remaining columns. If* $\mathbf{H}_2$ *contains exactly* $p - q$ *columns then* $\mathbf{H} = \mathbf{H}_2$, *otherwise proceed to the next step.*
- **Step 3.** *Pick one of the* $p \times (p - q)$ *submatrices of* $\mathbf{H}_2$, *denoted by* $\mathbf{H}_3$ *then apply the algorithm for matrix factorization presented in Sect. 6.2, since the reduced minors of* $\mathbf{H}_3$ *are guaranteed to be zero-coprime. Assume that the factorization yields* $\mathbf{H}_3 = \mathbf{H}_4 \mathbf{E}$, *for some polynomial matrix* $\mathbf{E}$. *Then, set* $\mathbf{H} = \mathbf{H}_4$ *and terminate.*

Example 6.5. Given a matrix $\mathbf{A} = [z_1^2 z_2^2 + 1 \ z_1^2 z_3 + 1 \ z_1 z_2^2 z_3]$, the goal is to compute a 3×3 unimodular matrix whose first row is identical to $\mathbf{A}$.

First, by using the program SINGULAR [82], a Gröbner basis of the module generated by columns of $\mathbf{A}$, with respect to the degree reverse lexicographical ordering $z_1 \succ z_2 \succ z_3$, is $\{1\}$ and a 3×1 polynomial matrix $\mathbf{B}$ such that $\mathbf{AB} = 1$ is $\mathbf{B} = [1 \ -z_1^2 z_2^2 \ z_1^3]^T$. In Step 2, the generating matrix of $Syz(\mathbf{A})$ is $\mathbf{H}_1 = [\mathbf{h}_1 \ \mathbf{h}_2 \ \mathbf{h}_3]$, where, $\mathbf{h}_1 = [z_1^2 z_3^2 + z_3 \ -z_3 \ -z_1^3 z_3 - z_1]^T$, $\mathbf{h}_2 = [z_1 z_2^2 z_3 \ 0 \ -z_1^2 z_2^2 - 1]^T$, and $\mathbf{h}_3 = [z_1^2 z_3 + 1 \ -z_1^2 z_2^2 - 1 \ 0]^T$.

Since h_1, h_2, h_3 are linearly independent, $H_2 = H_1$. It can be verified that none of the three set of two columns of H_1 can be used to form a MRP matrix. By using the n-D primitive factorization algorithm [83] on submatrix H_3 formed by the first two columns of H_2, the following factor H can be extracted, $H_3 \overset{\Delta}{=} [\ h_1 \ \ h_2\ ] = H_4 E$, where,

$$H \overset{\Delta}{=} H_4 = \begin{bmatrix} z_1^4 z_2^2 z_3 + z_1^2 z_3 + 1 & z_1^3 z_2^2 z_3 \\ -z_1^2 z_2^2 - 1 & -z_1 z_2^2 z_3 \\ -z_1^3(z_1^2 z_2^2 + 1) & -z_1^4 z_2^2 z_3 + 1 \end{bmatrix}, \quad E = \begin{bmatrix} z_3 & z_1 z_2^2 z_3 \\ -z_1 & -z_1^2 z_2^2 - 1 \end{bmatrix}.$$

The matrix H is indeed MRP and $AH = 0_{1 \times 2}$. By Proposition 6.1, H is a globally minimum generating matrix of $Syz(A)$. The unimodular matrix C and its associated inverse $\bar{A}$ are $C = (B \mid H)$, and

$$\bar{A} = C^{-1}$$
$$= \begin{bmatrix} z_1^2 z_2^2 + 1 & z_1^2 z_3 + 1 & z_1 z_2^2 z_3 \\ z_1^6 z_2^4 z_3 - z_1^2 z_2^2 + z_1^4 z_2^2 z_3 & z_1^6 z_2^2 z_3^2 + z_1^4 z_2^2 z_3 - 1 & z_1^5 z_2^4 z_3^2 - z_1 z_2^2 z_3 \\ -z_1^7 z_2^4 - 2z_1^5 z_2^2 - z_1^3 & -z_1^7 z_2^2 z_3 - z_1^5 z_2^2 - z_1^5 z_3 - 2z_1^3 & 1 - z_1^6 z_2^4 z_3 - z_1^4 z_2^2 z_3 \end{bmatrix}.$$

Appendix: Additional References Related to Conjecture 6.1

Note that Conjecture 6.1 was also posed as "A generalization of Serre's conjecture" in a joint paper by Z. Lin and N. K. Bose in the following paper:

Z. Lin and N. K. Bose, "A generalization of Serre's conjecture and some related issues," *Linear Algebra and Its Applications*, Vol. 338, pp. 125–138, Nov. 2001.

Solutions to Conjecture 6.1, using different methods, have been presented in the following papers:

[A1] J. F. Pommaret, "Solving Bose conjecture on linear multidimensional systems," in *Proceedings of the European Control Conference*, pp. 1853–1855, September 2001.
[A2] V. Srinivas, "A generalized Serre problem," *J. Algebra*, vol. 278, pp. 621–627, Aug. 2004.
[A3] M. Wang and D. Feng, "On Lin-Bose problem," *Linear Algebra and Its Applications*, Vol. 390, pp. 279–285, Oct. 2004.
[A4] A. Fabiánska and A. Quadrat, "Applications of the Quillen-Suslin Theorem to Multidimensional Systems Theory," *INRIA Rep.* 6126, 2007.
[A5] J. Liu, D. Li, L. Zheng, "The Lin-Bose Problem," *IEEE Transactions on Circuits and Systems II: Express Briefs*, Vol. 61, pp. 41–43, 2014.

References

1. W. K. Jenkins, "CAS Image Processing," in *A Short History of Circuits and Systems*, Franco Maloberti and Anthony C. Davies, Eds. 2016, River Publishers.
2. N. K. Bose, *Applied Multidimensional Systems Theory*, Van Nostrand Reinhold, New York, 1982.
3. D. E. Dudgeon and R. M. Mersereau, *Multidimensional Digital Signal Processing*, Prentice-Hall, 1984.
4. N. K. Bose, *Digital Filters: Theory and Applications*, Elsevier Science Publishing Co., Inc. North-Holland, Amsterdam, 1985.
5. N. K. Bose and P. Liang, *Neural Network Fundamentals with Graphs, Algorithms and Applications*, McGraw-Hill Book Company, NY, 1996.
6. N. K. Bose, *Multidimensional Systems Theory and Applications*, D. Reidel Publishing Company, Dordrecht, Holland, 2004.
7. N. K. Bose, "Multidimensional digital signal processing: problems, progress, and future scopes," *Proc. IEEE*, vol. 78, pp. 590–597, 1990.
8. D. Asimov, "There's no space like home," *The Sciences*, vol. 35, no. 5, pp. 20–25, 1995.
9. W. Fulton, *Algebraic Curves: An Introduction to Algebraic Geometry*, Benjamin/Cummings, Massachusetts, 1969.
10. H. Whitney, "Elementary structure of real algebraic variety," *Annals. of Math.*, vol. 66, pp. 545–556, 1957.
11. M. Nagata, "A general theory of algebraic geometry over Dedekind rings: ii," *Amer. J. Math*, vol. 80, pp. 380–420, 1958.
12. M. Auslander and D. A. Buschbaum, "Homological dimension in local rings," *Proc. Nat. Acad. Sci.*, vol. 45, pp. 733–734, 1959.
13. W. Rudin, *Function Theory in Polydiscs*, W. A. Benjamin Inc., New York, 1969.
14. B. L. Van der Waerden, *Modern Algebra*, vol. II, Ungar, NY, 1950.
15. E. I. Jury, *Inners and Stability of Dynamic Systems*, Krieger, Malabar, Florida, 1982.
16. S.G. Krantz, "What is several complex variables?," *American Mathematical Monthly*, pp. 236–256, 1987.
17. N. Levinson, "A polynomial canonical form of certain analytic functions of 2 variables at a critical point," *Bull. Amer. Math. Soc.*, vol. 66, pp. 366, 1960.
18. H. Whitney, *Differential and Combinatorial Topology*, Princeton Univ. Press, N. J., 1965.
19. N. Levinson, "Transformation of an analytic function of several complex variables to a canonical form," *Duke Math. Jour.*, vol. 28, pp. 345–354, 1961.
20. L. Bieberbach, *Analytische Fortsetzung*, Springer, Berlin, 1955.

© Springer International Publishing AG 2017   189
N.K. Bose, *Applied Multidimensional Systems Theory*,
DOI 10.1007/978-3-319-46825-9

21. K. V. Safonov, "Holomorphic extension of the two-dimensional Hadamard product," *Selecta Mathematica Sovietica*, vol. 8, no. 1, pp. 23–30, 1989.

22. T. Becker and V. Weispfenning, *Gröbner Bases:A Computational Approach to Commutative Algebra*, Springer-Verlag, New York, 1993.

23. S. V. Duzhin and S. V. Chmutov, "Gaydar's formula for the greatest common divisor of several polynomials," *Communications of the Moscow Mathematical Society*, pp. 171–172, 1993.

24. L. Bachmair and B. Buchberger, "A simplified proof of the characterization theorem for Gröbner bases," *ACM SIGSAM Bull.*, vol. 14, no. 4, pp. 29–34, 1980.

25. B. Buchberger, "A criterion for detecting unnecessary reductions in the construction of Gröbner bases," *Proc. EUROSAM 79, Marseille, Lecture Notes in Computer Science*, vol. 72, pp. 3–21, 1979, W. Ng (ed.).

26. B. Buchberger, "An algorithmical criterion for the solvability of algebraic systems of equations (German)," *Aequationes Mathematicae*, vol. 4, no. 3, pp. 374–383, 1970.

27. W. W. Adams and P. Loustanou, *An Introduction to Gröbner Bases*, vol. 3, Amer. Math. Society, Providence, R.I., 1994.

28. B. Buchberger and F. Winkler, Eds., *Gröbner Bases and Applications*, vol. 251 of *London Mathematical Society Lecture Notes Series*, Cambridge, 1998. Cambridge University Press, Proc. Intl. Conf. "33 Years of Gröbner Bases".

29. A. Seidenberg, "Constructions in algebra," *Trans. Amer. Math. Soc.*, vol. 197, pp. 273–313, 1974.

30. A. Tarski, *A Decision Method for Elementary Algebra and Geometry*, University of California Press, Berkeley, CA., 1930.

31. P. J. Cohen, "Decision procedures for real and p-adic fields," *Comm. Pure Appl. Math*, vol. 22, pp. 131–151, 1969.

32. A. Seidenberg, "A new decision method for elementary algebra," *Ann. Math.*, vol. 60, pp. 365–374, 1954.

33. N. Jacobson, *Lectures in Abstract Algebra*, vol. 3, Springer-Verlag, New York, 1964.

34. E. Engeler, *Introduction to the Theory of Computation*, Academic Press, New York, 1973.

35. J. M. Thomas, "Sturm's theorem for multiple roots," *Nat. Math. Mag.*, vol. 15, pp. 391–394, 1941.

36. B. D. O. Anderson, N. K. Bose, and E. I. Jury, "Output feedback stabilization and related problems - solution via decision methods," *IEEE Trans. Auto. Control*, vol. 20, pp. 53–66, Feb. 1975.

37. I. Yamada, K. Kurosawa, H. Hagesawa, and K. Sakaniwa, "Algebraic phase unwrapping and zero distribution of complex polynomials - characterization of multivariate stable polynomials," *IEEE Trans. Signal Process.*, vol. 98, no. 6, pp. 1639–1664, 1998.

38. I. Yamada and N.K. Bose, "Algebraic phase unwrapping and zero distribution of polynomials for continuous-time systems," *IEEE Trans. Circuits and Systems: I*, vol. 49, no. 3, pp. 298–304, 2002.

39. E. Zerz, *Topics in Multidimensional Linear Systems Theory*, Lecture Notes in Control and Information Sciences 256. Springer-Verlag London Limited, London, Great Britain, 2000.

40. P. A. Parrilo, "Semidefinite programming relaxations for semialgebraic problems," *Math. Prog. Ser. B*, vol. 96, no. 2, pp. 293–320, 2003.

41. C. N. Delzell, *A Constructive Continuous Solution to Hilbert's 17th Problem and Other Results in Semi-Algebraic Geometry*, Ph.D. thesis, Stanford University, Math. Dept., June 1980.

42. D. Henrion and A. Garulli (eds.), *Positive Polynomials in Control*, Springer-Verlag Berlin Heidelberg, 2005.

43. N. Z. Shor, "Class of global minimum bounds of polynomial functions," *Cybernetics*, vol. 23, pp. 731–734, 1987.

44. Bose, N., and C. Li. "A quadratic form representation of polynomials of several variables and its applications." IEEE Transactions on Automatic Control 13.4 (1968): 447–448.

45. Choi, Man-Duen, Tsit Yuen Lam, and Bruce Reznick. "Sums of squares of real polynomials." Proceedings of Symposia in Pure mathematics. Vol. 58. American Mathematical Society, 1995.

46. Parrilo, P. A. (2000) Structured semidefinite programs and semialgebraic geometry methods in robustness and optimization (Doctoral dissertation, California Institute of Technology).
47. P. Parrilo, "Exploiting algebraic structure in sum of squares programs," in *Positive polynomials in control*, D. Henrion and A. Garulli, Eds., pp. 181–194. Springer, 2005.
48. Y. A. Brychkov, H. J. Glaeske, A. P. Prudnikov, and V. K. Tuan, *Multidimensional Integral Transformations*, Memoirs Amer. Math. Soc. Gordon and Breach Science Publishers, Philadelphia, 1992.
49. H. Miyakawa, "Sampling theorem of stationary stochastic variables in multidimensional space," *J. Inst. Elec. Commun. Engrs. (Japan)*, vol. 42, pp. 421–427, 1959.
50. D. P. Peterson and D. Middleton, "Sampling and reconstruction of wavenumber-limited function in n-dimensional Euclidean space," *Information Control*, vol. 5, pp. 279–323, 1962.
51. R. M. Mersereau, "The processing of hexagonally sampled two-dimensional signals," *Proc. IEEE*, vol. 67, pp. 930–949, 1979.
52. J. W. Goodman, *Introduction to Fourier Optics*, McGraw-Hill, New York, 1968.
53. A. H. Zemanian, *Distribution Theory and Transform Analysis*, Dover Publications Inc., New York, 1965.
54. M. J. Lighthill, *Introduction to Fourier Analysis and Generalized Functions*, Dover Publications Inc., Cambridge, England, 1960.
55. E. Dubois, "Motion-compensated filtering of time-varying images," *Multidimensional Systems and Signal Processing*, vol. 3, pp. 211–239, May 1992.
56. A. Cohen and I. Daubechies, "Non-separable bidimensional wavelet bases," *Revista Matematica Iberoamericana*, vol. 9, pp. 51–137, 1993.
57. N. K. Bose and E. I. Jury, "Positivity and stability tests for multidimensional filters (discrete-continuous)," *IEEE Trans. Acoust., Speech, and Signal Process.*, vol. 22, pp. 174–180, June 1974.
58. H. G. Feichtinger and K. Groechenig, "Iterative reconstruction of multivariate bandlimited functions from irregular sampling values," *SIAM J. Math. Anal.*, vol. 23, pp. 244–261, 1992.
59. Gilbert G. Walter, "A sampling theorem for wavelet subspaces," *IEEE Trans. Inform. Theory*, vol. 38, no. 2, pp. 881–884, March 1992.
60. Alfred Fettweis and Sankar Basu, "Multidimensional causality and passivity of linear and nonlinear systems arising from physics," *Multidimensional Systems and Signal Processing*, vol. 22, pp. 5–25, Mar. 2011.
61. Rikus Eising, *2-D systems: An algebraic approach*, Ph.D. thesis, Mathematisch Centrum, Amsterdam, 1979, https://pure.tue.nl/ws/files/2320112/159120.pdf.
62. J. H. Justice and J. L. Shanks, "Stability criteria for n-dimensional filters," *IEEE Trans. Automatic Control*, vol. 18, pp. 284–286, June 1973.
63. M. Vidyasagar and N. K. Bose, "Input output stability of linear systems defined over measure spaces," in *Proc. 18th Midwest Symp. Circuits and Systems*, Montreal, Canada, 1975, pp. 394–397.
64. D. M. Goodman, "Some stability properties of two-dimensional linear shift-invariant digital filters," *IEEE Trans. Circuits and Systems: I*, vol. 24, pp. 201–208, 1977.
65. Z. Lin and N. K. Bose, "A generalization of Serre's conjecture and some related issues," *Linear Algebra and Its Applications*, vol. 338, pp. 128–138, 2001.
66. M. Scheicher, "A generalization of jury's conjecture to arbitrary dimensions and its proof," *Mathematics of Controls, Signals and Systems*, vol. 20, pp. 305–319, 2008.
67. M. Scheicher and U. Oberst, "Multidimensional BIBO stability and jury's conjecture," *Mathematics of Controls, Signals and Systems*, vol. 20, pp. 81–109, 2008.
68. M. Scheicher and U. Oberst, "Proper stabilisation of multidimensional input/Output behaviours," *Multidimensional Systems and Signal Process.*, vol. 20, pp. 185–216, 2009.
69. F. G. Boese and W. J. Luther, "Enclosure of the zero-set of polynomials in several complex variables," *Multidimensional Systems and Signal Process.*, vol. 12, pp. 165–197, 2001.
70. Yuval Bistritz, "Stability testing of two-dimensional discrete-time systems by a scattering-type stability table and its telepolation," *Multidimensional Systems and Signal Process.*, vol. 13, pp. 55–77, 2002.

71. Minoru Yamada, Li Xu, and Osami Saito, "Further results on Bose's 2-D stability test," in *Proc. Fourteenth Int. Symp. Mathematical Theory of Networks and Systems (MTNS-2000)*, Perpignan, France, June 2000.

72. Sh. A. Dautov, "On absolute convergence of the series of taylor coefficients of a rational function of two variables: Stability of two-dimensional recursive digital filters," *Soviet Math. Dokl.*, vol. 23, no. 2, pp. 448–451, 1981, (American Mathematical Society Translations).

73. D. C. Youla, "Two observations regarding first-quadrant causal BIBO-stable digital filters," *Proc. IEEE*, vol. 78, pp. 598–603, 1990.

74. Andrew T. Tomerlin and William W. Edmonson, "BIBO stability on D-dimensional filters," *Multidimensional Systems and Signal Process.*, vol. 13, no. 3, pp. 333–340, 2002.

75. J. H. McClellan, T. W. Parks, and L. R. Rabiner, "A computer program for designing optimal fir linear phase digital filter," *IEEE Trans. Audio Electroacoustics*, vol. AU-21, pp. 506–526, 1973.

76. T. S. Huang (ed.), *Topics in Applied Physics*, vol. 42, Springer-Verlag Publishers, 1981.

77. R. M. Mersereau, W. F. G. Mecklenbräuker, and T. F. Quatieri, "McClellan transformations for two-dimensional digital filtering: I - Design," *IEEE Trans. Circuits Systems*, vol. CAS-23, pp. 405–414, 1976.

78. W. K. Jenkins, J. C. Strait, and R. P. Faust, "Convergence properties of a constrained 2-D adaptive digital filter," in *Proc. 22nd Asilomar Conference on Signals, Systems, and Computer*, Pacific Grove, CA, Nov. 1988, pp. 250–254 (invited).

79. J. L. Shanks, S. Treitel, and J. H. Justice, "Stability and synthesis of two dimensional recursive filters," *IEEE Trans. Audio Electroacoustics*, vol. AU-20, pp. 115–128, June 1972.

80. S. Treitel and J. L. Shanks, "The design of multistage separable planar filters," *IEEE Trans. Geoscience Electronics*, vol. 9, pp. 10–27, 1971.

81. J. P. Guiver and N. K. Bose, "Polynomial matrix primitive factorization over arbitrary coefficient field and related results," *IEEE Trans. Circuits and Systems: I*, vol. 29, pp. 649–657, 1982.

82. G.M. Greuel, G. Pfister, and H. Schoenemann, *Singular version 1.2 User Manual*, Centre for Computer Algebra, University of Kaiserslautern, Germany, June 1998.

83. Z. Lin, "Notes on n-D polynomial matrix factorization," *Multidimensional Systems and Signal Process.*, vol. 10, pp. 379–393, 1999.

84. T.Y. Lam, *Serre's Problem on Projective Modules*, Springer, Berlin, springer monographs in mathematics, edition, 2006.

85. Z. Lin, "On syzygy modules for polynomial matrices," *Linear Algebra and Its Applications*, vol. 298, pp. 73–86, 1999.

Printed in the United States
By Bookmasters

Printed in the United States
By Bookmasters